T0320357

DECODING
CHINA'S CAR
INDUSTRY

40 YEARS

DECODING CHINA'S CAR INDUSTRY

40 YEARS

LI Anding

Translated by
Chris R Lanzit, Graeme Ford

World Scientific

W JERSEY · LONDON · SINGAPORE · BEIJING · SHANGHAI · HONG KONG · TAIPEI · CHENNAI · TOKYO

Published by

World Scientific Publishing Co. Pte. Ltd.

5 Toh Tuck Link, Singapore 596224

USA office: 27 Warren Street, Suite 401-402, Hackensack, NJ 07601

UK office: 57 Shelton Street, Covent Garden, London WC2H 9HE

Library of Congress Cataloging-in-Publication Data
Names: Li, Anding, author. | Lanzit, Chris R., translator. | Ford, Graeme (Translator), translator.
Title: Decoding China's car industry : 40 Years / Anding Li ; translated by Chris R Lanzit,
      Graeme Ford.
Other titles: Che ji. English
Description: New Jersey : World Scientific, [2022]
Identifiers: LCCN 2021029484 | ISBN 9789811241109 (hardback) |
      ISBN 9789811241116 (ebook) | ISBN 9789811241123 (ebook other)
Subjects: LCSH: Automobile industry and trade--China.
Classification: LCC HD9710.C52 L5313 2022 | DDC 338.4/76292220951--dc23
LC record available at https://lccn.loc.gov/2021029484

British Library Cataloguing-in-Publication Data
A catalogue record for this book is available from the British Library.

车记: 亲历 • 轿车中国三十年
Originally published in Chinese by SDX Joint Publishing Co. Ltd., 2017

For any available supplementary material, please visit
https://www.worldscientific.com/worldscibooks/10.1142/12391#t=suppl

Desk Editor: Lixi Dong

Typeset by Stallion Press
Email: enquiries@stallionpress.com

Printed in Singapore

Decoding China's Car Industry: 40 Years

Written by Li Anding
Translated by Chris R. Lanzit and Graeme Ford

Contents

Preface

It Seems a Lifetime Ago

This book is a record of the years 1978 to 2018, a period when China's car industry was built from scratch and moved rapidly to become a global phenomenon in production and sales. I participated in and witnessed the main events and details of this history.

If China had not opened up, global auto industry funds and technology would not have been brought in. Chinese companies, especially private ones, would not have entered into global cooperation and competition. And China would still be a bicycle kingdom today.

Chinese automobiles kept growing at double-digit rates for many years. That came to an abrupt end in 2018, closing the circle in my forty year record. We can't tell whether China will go from being an automobile power to be a great power, or go into a period of declining growth or even downturn in the future.

The automobile era began 130 years ago. China was the last to enter, but it progressed the fastest. It was not even 40 years since joint venture automobile production was tried, only 30 years since the decision to establish a car industry, and just 20 years since the government gave approval for families to own cars.

Are today's young Chinese people aware that more than half a century ago, private cars and even car production were strictly banned as they were seen as being capitalist? Also, as China had erected the world's highest tariff barriers, total production was only 3,000 vehicles per year until the 1980s, not even near to a day's production in a foreign factory?

All these were the starting points of the story of Chinese cars 40 years ago. If you told Chinese people then that their children would enjoy the right to have a car culture, and that China would become first in automobile production and sales, surpassing the US — the kingdom of cars, they would think it was a dream. Forty years is not a long time, but looking back, it seems a lifetime ago.

Things have changed so fast that there hasn't been time to take it all in and reflect on it. Someone has to sort it all out, from the beginning until now.

The starting point of my journalistic career coincided with the beginning of the Chinese auto industry. I moved among its decision-makers and principal enterprises. I personally experienced the whole process and many of the key events of the past 40 years. I gained a unique perspective and an understanding that no outsider could get. I have been very lucky.

Chinese car founder Rao Bin said to me in 1982 that today's news is tomorrow's history. I have always kept this in mind. Writing this history is something I have to do during my lifetime.

I believe it was the farsightedness of decision-makers and governing departments which, starting from scratch, brought about the rapid development of the car industry in China. At the same time, the mutual constraints between these departments, and their repeated mistakes, caused one complication after another. We were winning and losing at the same time. However, the automobile industry changed from an exemplar of the planned economy to the most market-oriented key industry in China. It saved face for China's economy, even at the most turbulent moments of the global financial crisis.

Nevertheless, tangles run right through the 40-year ups and downs of the Chinese car industry.

Do we close our country off and be self-reliant, or do we seize the opportunities of openness and embrace globalization? In joint ventures, do we struggle over everything, dreaming different dreams in the same bed, or do we cooperate and mutually prevail, generating local technology and the ability to develop? Do we build the walls of state-owned enterprises higher and kill off grassroots competitors, or do we open the door and give independent brands the right to survive and compete? Are we satisfied with the protected status of Chinese cars, or do we encourage Chinese brands to become strong in a competitive environment in accordance with the WTO rules? Do we pursue the leading technologies of new energy, seeking breakthroughs as we learn, or do we deceive ourselves by trying to overtake on a curve? History carries a lot of silt along with it as it rolls forward.

What is a car? In China, cars are politics, technology, the economy, rights, wealth, culture, and loved ones. They are the fleet footed runners that travel a hundred times further, a rolling stream of social wealth, and a classic product of globalization. They are the spiritual sustenance for populism and the chief culprit of pollution and congestion. They are the leading lights of new energy around the world and the embodiment of indigenous innovation of Chinese brands. They are symbols of the disparity between rich and poor and the antagonism between a government and its people.

China has encouraged private consumption of cars for less than 20 years, but they are suddenly appearing among us at a rate of more than 20 million vehicles a year. Chinese people still don't know whether to love them or hate them, use them or curse them, encourage them or limit them.

This book chronicles the tangled stories of events, companies, and characters in different periods, with a chapter on each subject. This is a private version of the history of Chinese cars. It seeks to restore the truth by speaking freely, without the balance and restraints which will face future official writers.

I have been able to complete this book with encouragement and support from the Chinese and foreign car industries, and seniors, friends, and young people in the media. They provided valuable memories, judgments, historical materials, and pictures during the writing process. The list is too long, so I will not list all of their names. I express my deep gratitude to them.

Finally, I want to thank World Scientific Publishing Company for selecting my book among the many publications in China to be translated into English so that readers around the world can share what

I have seen, thought, and felt. I want the world to know and recognize the talent and creativity that ordinary Chinese people have shown in the past 40 years. They have dreamed of and hoped for cars that are more intelligent, cleaner, safer, and more comfortable, and now can own them like people in other countries.

Li Anding

About the Author

Li Anding is an automotive columnist for various newspapers and websites and a former editorial board member and senior journalist at Xinhua News Agency. He has been writing about the Chinese car industry since the 1980s and is known as the "leading media reporter on cars". He is also the author of *Thousand Hands and Thousand Eyes: On the Stage and Behind the Curtain of Chinese Reformation, The Lure of Family Owned Cars in China*, and *Notes of Cars Outdoors*.

About the Translators

Chris R Lanzit is a certified professional translator accredited since 2017 by Australia's National Accreditation Authority for Translators and Interpreters (NAATI). He has lived and worked in China for over 20 years. He served in the US Air Force from 1974–1995 and was posted to Beijing as Assistant Air Attache from 1990–1992. He subsequently worked in Hong Kong and Beijing for Hughes Space & Communications and the Boeing Company. From 2004 to 2012 he represented several US standards developers in Beijing. He is a keen reader of Chinese history. His first major translation was a multi-part documentary on Yunnan's 1930s Warlord-Governor Long Yun.

Dr Graeme Ford is a NAATI accredited Chinese translator. He has taught Chinese language and translating at colleges and universities in Australia. He has lived, studied and worked in several cities in China. He has a strong interest in comparative Chinese and English grammar. His PhD research is in the history of translating at the Ming Court. He has collaborated with Chris Lanzit on several translating projects.

Chapter 1

Crossing the River by Feeling the Stones

One cloudy afternoon just after Spring Festival in 1982, I had an appointment at the First Ministry of Machinery Industry located in a group of buildings with green glazed tile roofs in Sanlihe in Beijing. I happened to meet Minister Rao Bin in the elevator.

"We've just been approved by the State Council to establish an automobile industry corporation. I've asked a few comrades to discuss it. Please come along and listen," Rao Bin said as he stepped out of the elevator. "Today's news is tomorrow's history. Isn't that what you journalists say?"

Rao Bin was one of the founders of the Chinese automobile industry. At that moment in history, he wore a pair of old cloth shoes and a blue-gray Sun Yat-sen tunic. He was tall and powerful, and he combed his gray hair in a characteristically slicked back way. He always cut a fine figure, no matter how simply he dressed or how his fates changed.

I was deeply impressed by what he said, and it inspired a sense of mission in me to use journalism to chronicle automobiles in China. This sense of mission accompanied my whole 40-year career in journalism.

A series of structural adjustments began in agriculture and light and heavy industries in the late 1970s. The automobile industry was at a low point and faced a difficult turnaround because it was categorized as a heavy industry.

The Reform and Openness[1] policy began in the 1980s. Establishing the car industry was on the agenda of the newly created China National Automotive Industry Corporation (CNAIC) from the beginning. However, it met with firm opposition from leading authorities at the time who said, "We can't give cars a green light, because they are products of the bourgeois lifestyle."

However, Deng Xiaoping, the chief architect of Reform, supported the start-up of the car industry at this critical moment by pointing out a creative new path — joint ventures for automobiles. This opened a way for an auto industry which was facing serious shortages of funds and technology.

1. A pathfinder's solemn mission

Today's news is tomorrow's history

I became a journalist at Xinhua News Agency at the end of 1978. It was the watershed between the Cultural Revolution and a new era. The Third Plenary Session of the 11th Congress of the Communist Party of China (CPC) had just begun. In my 40-year career as a journalist since then, I have become the recorder of the whole course of Reform and Openness.

People have become used to the term, but at the time there was a sense of rebirth for people who had just been through the Cultural Revolution and class struggle. Things blew hot and cold, and there were fierce contests between old and new ideas.

In 1980, I was assigned to cover the manufacturing sector, mainly the machinery, textile, and military industries. Only a few media outlets could interview the central ministries and there weren't many

---

[1] The Chinese economic reform known in the West as the Opening of China refers to the program of economic reforms termed "Socialism with Chinese characteristics" and "socialist market economy" in the People's Republic of China which reformists within the Communist Party of China — led by Deng Xiaoping — started on 18 December 1978. Before the reforms, the Chinese economy was dominated by state ownership and central planning.

young journalists at that time. The official atmosphere was still open and there was none of the red tape which exists today. I had a permanent pass for each ministry and a canteen meal ticket. I went from one office to the next and everyone was friendly, from the minister to the ordinary staff. I chatted, read documents, got news tips, and then ran back to my office. A lot of scoops came my way every day. It was exciting.

Rao Bin was the Minister of Machinery Industry. He had just been transferred back to Beijing from China's Second Automobile Factory in 1981.

He was a legend as the founder of the Chinese automotive industry. I was often in his office. Gu Yutian, his secretary, always called me in to sit for a while whenever the outside door was open. When the minister had something for him to do, Gu would send word that Anding was there. Minister Rao would then ask me in for a chat or sometimes for a formal interview.

There were more than 120 automobile factories in China at that time, belonging to different industries including machinery and communications and to municipal administrations. Most of them were small local factories with small scale production. Only 175,000 vehicles were produced nationally each year. Their technology and quality were very poor. It wasn't uncommon for a driver to be trapped on the road if he couldn't turn his hand to repairing his car.

I followed Minister Rao into the conference room that early spring afternoon in 1982. Several of the leaders were waiting, including Liu Shouhua, Feng Ke, and Hu Liang of the National Automobile Bureau. The Minister took out a document marked with the State Council's red seal and said, "I am delighted to tell you that the Central Government has decided to establish a China National Automotive Industry Corporation to fix up the shortcomings in the automobile industry. This will be a trial, providing experience for reforms and structural adjustments to other industries. The industry is facing major managerial reforms, so the corporation will be under the direct leadership of the State Council." He also revealed that the Central Government had confirmed Li Gang as General Manager.

I knew Li Gang. He was still the director of First Auto Works (FAW) in Changchun at the time. I didn't know about his new appointment. He impressed me at an annual National Machinery Industry Conference by not attending a movie in the evening but hiding in his room and quietly reading instead.

I had a premonition on the day the State Council issued the document that Rao Bin would leave his position as head of China's Machinery Industry to throw himself back into the ups and downs of the auto industry.

7 May 1982 was set for the establishment of the CNAIC.

On the afternoon of the last day of the meeting, Bo Yibo, Vice Premier and Director of the National Machinery Industry Committee, came to the venue to announce a decision just made by the Secretariat of the CPC Central Committee and the State Council. I remember the atmosphere at the conference was rather tense. Bo Yibo announced that instead of being a controlling corporation as in the original document, CNAIC would be a bureau-level institution under the Ministry of Machinery Industry, and not directly under the State Council. Rao Bin would serve as chairman of the corporation, Li Gang as general manager, and Chen Zutao as chief engineer.

I noticed that Rao Bin's face remained calm and the audience was solemn and respectful. The status of the department head would directly affect the development of an industry. Rao Bin and the automotive industry leaders and experts present were very aware of this Chinese peculiarity. Subsequent events proved their concerns were not unfounded. The old system was convoluted and complex. CNAIC had made a breakthrough, but it was hard to consolidate this progress and it ended unhappily.

Removing the word *General* from China National Automotive Industry Corporation

The economy was in the middle of a major adjustment around the time CNAIC was established. The State Planning Commission, the economic decision-making department, listed cars under projects that would promote energy conservation by limiting production and impounding them. Development of the automobile industry would clearly be constrained by these policy countermeasures. The company was caught at a time when the State was planning to cut production just as it was being set up. Planned production was reduced to 80,000 units in 1982, only one-third of the 1980 output. A large number of private vehicles were ordered off the road and put in storage.

Rao Bin was deeply humiliated. He personally drafted a report to the Central Committee of the CPC, stating, "Countries all over the world, oil-producing and oil-importing, are developing automobile transportation. This is determined by the laws of economic development." He suggested the State to construct roads, increase asphalt paving, and eliminate restrictions on automobile production.

A week later, on 7 November 1982, the Premier of the State Council endorsed the report on "A Series of Major Technical and Economic Policy Issues Concerning Transportation" and tasked the State Planning Commission with demonstrating its feasibility. "As for restricting production, as long as there is a market, of course production is to be allowed and mustn't be restricted."

The automobile market was in a slump when the corporation was first formed. A Dongfeng 5-ton truck sold for only 18,000 yuan and the manufacturer had to provide financing. The people working in the corporation were the first generation of entrepreneurs. They had broad vision, a wealth of experience, and didn't blame others. They put their minds to bringing the Chinese auto industry out of its predicament.

To develop the automobile transportation market, the company decided in the early spring of 1983 to advise the Central Government to solve coal transportation problems by providing heavy-tonnage vehicles. Small coal mines had sprung up in Shanxi as rural economic reforms were promoted and rail capacity was limited. Large amounts of coal couldn't be transported out. It was piled up in the mountains and lain exposed to wind and rain, sometimes even spontaneously combusting.

CNAIC General Manager Li Gang decided to go to Datong in person to inspect resources, routes, capacity, costs, and other issues of coal transport. He invited me to go along. A Red Flag van produced by

FAW carried Li Gang, his secretary Zhang Ning, and myself. We left Beijing early in the morning, heading west through Zhangjiakou Gate.

Li Gang graduated from Tsinghua University in 1948 with a major in automobile manufacturing. He went to the Soviet Union in 1952 to join an FAW Working Group at the Ministry of Heavy Industry. He returned in 1953 to head up the technical department at FAW's engine branch. In 1965, he was Deputy Chief Engineer. He became FAW's Deputy Factory Manager and and then Factory Manager after the Cultural Revolution.

It was dark when we arrived in Datong. Li Gang asked the driver to drive the car directly to the local transport company and we went to a room upstairs. Shortly after we settled down, the director of the Municipal Party Committee rushed in, saying that the city leaders had arranged a welcome banquet at a hotel, and because Li was an alternate member of the CPC Central Committee, he should move to the hotel where it would be easier to guard him. Li Gang firmly refused and the director insisted again. Li Gang said with some heat, "You go back. I'll stay here where it's easier to chat with the cadres and drivers who manage transportation."

When he ended his investigations that evening, this high-level Central Government executive, his secretary, and a reporter slept in a simple room without a bathroom. That was quite normal in that era.

After staying in Datong for a night, we assembled with two heavy trucks and headed to the small coal pits in the mountains. The wind beyond the Great Wall was very sharp and cold. Li Gang stopped the vehicle from time to time, jumping down and pacing out the width of the road to calculate the amount of traffic it could carry. Close to the coal pits, the road simply disappeared, and the vehicles pushed forward along a rutted track. We caught up with an old Jiefang (Liberation model) truck. He stopped it and got the driver, who wore a black fur coat, to squeeze into our vehicle for a conversation. Later, the truck driver asked me, "What is this guy up to? He sure knows a lot about automobiles!"

Li Gang went into one small coal pit after another to find out about the primary cost of coal and the freight and road conditions. Shanxi's coal truck transport business flourished and became a showcase for how vehicles could promote economic growth.

Li Gang was elected as an alternate member of the Central Committee at the CPC's 12th National Congress in September 1982. Recalling his term in office, he was very proud to have improved the strategic position of the automobile industry in the national economy. With the active appeal of CNAIC, the automobile industry was listed as a pillar of production for the first time at the Third Plenary Session of the 12th Party Congress.

I spent one or two days at the corporation almost every week during that period. Fortunately for me, Rao Bin, Li Gang, and a group of outstanding car experts treated me as a friend. They were my first teachers as I got involved in car reports and reviews. They talked with me about the great role that car transport played in the development of the world economy, and about the car era that had been delayed in China. That excited me and, in the spring of 1983, I wrote a special article for Xinhua News Agency's *Outlook Weekly* magazine, "Automobile Age on the Horizon," attacking the limits on automobile production.

In the article, I said that we should not just focus on freight vehicles but also on passenger cars that had its demand suppressed. Although China's conditions were different from those of other countries, it was also unscientific to adopt a static, one-sided view of production of medium-sized buses, cars, and motorcycles.

The original intention of establishing the company was to set up a trust and carry out reform. It was to form an economic entity having the management functions of the automobile industry, operating independently of production, operations, and budget, in accordance with the requirements of the central decision-making level under the guidance of a national unified plan.

The automobile industry tried to set up leading backbone enterprises in 1982 to resolve the disorder left over from the past. Six jointly managed companies were established — Jiefang (Liberation), Dongfeng, Beijing-Tianjin-Hebei, Heavy-duty, Nanjing, and Shanghai. There was both competition and cooperation among them.

CNAIC also organized secondary companies for investment, sales, import and export, parts, and supply of materials. The China Automotive Technology Center was also established in Tianjin.

For 30 years after the establishment of New China, all automobiles, as production items, were allocated under the state plan. Markets were quite unknown to the auto industry. However, it was Chinese farmers who first made cars a commodity in 1984. I wrote a newsletter that year, entitled "New Customers for Automakers," about peasants in Henan and Shanxi who, eager to make money from rapid transport, went to Nanjing Automobile Factory with bundles of banknotes in their waistbands to buy light trucks.

Li Gang, General Manager of China National Automotive Industry Corporation and a specialist in engines, explaining the rotary engine developed by FAW to Marshal He Long and General Luo Ruiqing in 1961. Only a few companies in the world had that technology then.

Hu Yaobang inspecting Second Auto Works in June 1982. Chief Engineer Chen Qingtai showed him around the vehicle laboratory.

First Auto Works ended 30 years of making Liberation Trucks in 1982 and developed a new generation of cars. Group photo at the test track with author first from right.

The auto industry grew from a state of recession to a degree of initial prosperity by going to the market. Automobile production was 175,000 in 1981 and 440,000 in 1985. It increased one-and-a-half times in four years and by 25 percent each year. Output increased from 6.9 billion yuan to 25.9 billion yuan, nearly quadrupling in four years, with annual growth of 39 percent. This reversed losses made when the company was first set up.

One noteworthy thing that Rao Bin did was to direct the industry to bring itself up to contemporary international standards.

He traveled to major automobile companies all over the world to compare and choose. The Austrian Steyr heavy-duty vehicle series and the US Cummins engine were brought in and supplied to heavy-duty automobile groups in Jinan in Shandong, in Baoji in Shaanxi, and in Dazu in Sichuan.

While foreign trade departments were importing a lot of light-duty vehicles, all the drawings and materials for Japanese Isuzu N-series diesel engines were obtained free of charge. They were used for a national upgrade of light vehicles. This created a precedent of exchanging markets for technology. The introduction of the Iveco light vehicle series from Fiat Italy put Nanjing Automobile Factory's Yuejin 30 years ahead. The Japanese Daihatsu mini-van was introduced in Tianjin.

At the ceremony for the introduction of the Iveco light vehicle, I asked Chairman Rao for the standards of a world car in the 1980s.

"What's being put into production are mainstream models which can keep being developed with realistic and potential markets," he answered. "Their introduction is dynamic, including dynamic technological development of counterpart products during the contract period. We can't keep using the same technology for 30 years anymore."

The foresight of first-generation automotive industry leaders could be seen in his statement.

A series of heavy, medium, light, and micro trucks and engines with 1980s standards were finally established on Chinese soil. Passenger cars were still restricted at that time.

Earthquake at CNAIC and demise of the Red Flag

Funds were extremely tight as the economy recovered in the 1980s and investment reform was also a priority. Before the establishment of CNAIC, the government had already determined that no investment would be made to reform FAW. The construction of Second Auto Works (SAW) was not yet complete and a further two to three hundred million yuan was needed. State investment for developing light and heavy vehicles was out of the question. Therefore, the corporation proposed the State to adopt such means as incremental interest-rate increases, return of funds for depreciation and major overhauls, and bank loans, to ensure that funding could be guaranteed for production, quality, development, and construction.

However, Rao Bin and Li Gang didn't foresee the trouble in their own backyard and an earthquake suddenly struck. The leaders of FAW wrote to the Central Committee criticizing the management of CNAIC in July 1984. The letter stated that a very large enterprise like FAW couldn't be made a workshop for the nation without giving it the right to develop. They demanded that FAW be directly listed in the state plan and have greater autonomy.

Viewed from today, this letter reflected the earliest consciousness of the need for loosening up of state-owned enterprises and it got the central leadership's attention. The Central Leading Group for Financial and Economic Affairs held a meeting about it.

Li Gang and I spoke about it some years later. "The Leading Group held a meeting at Beidaihe on 11 August 1984 and listened to FAW's report. Rao Bin and I, who still knew nothing about it, were called in as well. The leaders of the State Council, who were very tough in those days, demanded the results of CNAIC's reforms carried out over the past two years. FAW and SAW were to be given a separate listing in the State Planning Commission's plan," he said.

"The establishment of FAW and SAW as independent entities actually eroded CNAIC's authority and was a major setback for it. Chairman Rao and I were completely unprepared. I didn't sleep for a few nights," Li Gang said. "There was still fierce confrontation between reformists and conservatives. There were many policy changes in the Central Committee. Leaders were deciding major policy directions without proper thought and consideration. This was inevitable while we were still crossing the river by feeling the stones."

Another bigger and unexpected blow fell at that meeting — the order to stop producing the Red Flag sedan, which had been in production for 25 years.

For first generation automakers such as Rao Bin and Li Gang, the Red Flag had been the best thing in their lives.

Mao Zedong rode in the first Dongfeng sedan developed by FAW in the summer of 1958 when the Great Leap Forward was in full swing. Then the Central Government ordered it to develop a high-class sedan. Rao Bin, the Factory Director, immediately organized a large-scale campaign. The first Red Flag 770 limousine, draped in red and with drums beating, was driven to the Jilin Provincial Party Committee on 1 August.

When people speak of the first generation of Red Flag sedans today, they often say that it didn't conform to a national style. Then they said that the sedans were beaten out with sledgehammers. The two sentences contained both praise and criticism. I didn't know this was wrong until I chatted with Li Gang.

Li Gang, who had participated in the development of the Red Flag's engine, said that the V-8 engine was its bright spot but there were huge problems. "That was the best time of my life. We were working on engines for the Dong Feng, Red Flag, and an off-road vehicle all at the same time. I once went without sleep for 96 hours. I didn't care because I was in good health."

V8 engines were high technology in the 1950s. As well as being used in high end luxury cars in the US, they had just been used in the Chaika (Seagull) cars Soviet leaders rode in. V8 engines weren't used in Germany and Japan until years later.

A Red Flag 772 equipped with bullet-proof glass and thick armor plating was developed especially for national leaders. The Red Flag 771 with large rear seating, room for a translator, and an extra row of seats subsequently went into production. However, the factory only received limited funds from the State and nothing for follow-on development.

Red Flag production was entirely self-sufficient. Every component was produced domestically. But after 25 years, quality and performance were lagging further and further behind international standards.

Only a small number of Red Flags were produced, just 1,500 vehicles in 25 years. You couldn't produce on such a small scale without losing money. FAW saw production of the Red Flag as a glorious political task, so it relied on Jiefang trucks from 1958 to 1984.

The shortcomings of the Red Flag sedan were exposed as the horizons of its passengers became broader when China was opening up in the 1980s.

Li Gang recalled the 1984 meeting of the Central Leading Group for Financial and Economic Affairs where the Red Flag was discussed. "The head of the State Council turned to Rao Bin and said that it was fuel-intensive, slow, and unreliable. Just put a stop to it."

"Rao Bin argued that a four-seater sedan is not the same as a twelve-seater. The twelve-seater has much greater capacity than the four-seater. It is big and heavy. Naturally its fuel consumption is higher, but not much more than for similar foreign vehicles. I added that only one Red Flag sedan could be produced for the cost of ten Jiefang trucks, but the leaders in Zhongnanhai rode in it and it was a matter of patriotism for us."

"The leader said, 'Don't get yourself worked up.' Those were his words. 'Do what I say and put an end to it.'"

"Then what shall we do?" Rao Bin asked.

"We'll import them," he said.

"That's how he was turned down. Right to his face."

There were different versions of the end of the Red Flag. I recorded the memories of those who were there, word for word. They all seemed to agree.

This undoubtedly dealt a fatal blow to the Chinese auto industry, to CNAIC, and especially to Rao Bin and Li Gang. They witnessed the end of the car in a humiliating face-to-face encounter.

Automobile imports were highest in the year the Red Flag was stopped. A lot of Japanese Crown cars began to be imported as official vehicles. The Shanghai was the only domestic sedan left.

A last glorious memory of the first-generation Red Flag sedan was at the celebration for the 35th anniversary of the founding of the People's

Republic of China on 1 October 1984. Deng Xiaoping, China's top leader and chief architect of reform, rode in a specially built Red Flag convertible sedan to review the army, navy, and air forces in Tiananmen Square.

The breakaway of FAW and SAW and the demise of the Red Flag sedan doomed CNAIC's institutional reforms. The company convened an end-of-term meeting of its Board of Directors in Beijing in the summer of 1986. It was held in the same dignified atmosphere as three years before, with a speech by Bo Yibo who represented the highest decision-making level of the government. He announced that Rao Bin would be removed from the chairmanship and transferred to the Central Advisory Commission. Li Gang took over as Chairman and Chen Zutao as General Manager. Bo emphasized a major change from the original system where the chairman had responsibility to one where the general manager had responsibility. This is how Rao Bin and Li Gang were removed from the operational center of the company.

Mao Zedong inspecting FAW's Dongfeng car at Zhongnanhai in May 1958. He said emotionally, "Now we can ride in a car of our own."

The company was reorganized as the China National Automotive Federation (CNAF) on 29 June 1987. Chen Zutao was re-elected chairman.

2. Deng Xiaoping decides on automobile joint ventures

1978: We can set up joint ventures

When I interviewed Dr. Carl Hahn, the former chairman of Volkswagen (VW), something he said surprised me. He said that joint ventures were an American invention, but they did well in China.

Global car giant General Motors (GM) formed a joint venture in the US with Toyota in the 1970s called New United Motor Manufacturing, Inc. (NUMMI) to learn about Toyota's operations and management.

The GM Chairman led a delegation to China in October 1978. When discussing the introduction of heavy truck technology, he recommended that China should use joint ventures. "Simply put, joint ventures allow us to put our wallets together, run businesses together, make profits together, and pay for them together. Put plainly, joint ventures are like marriages, they build shared families."

Li Lanqing (a former Vice Premier), who was called to Beijing from SAW to take part in the negotiations, recalled, "What he said made sense, but I couldn't accept it emotionally. If one is a big capitalist, and another a communist, how can we get married?" Nevertheless, he put it in his report to the State Council.

When he saw the briefing, Gu Mu, the Deputy Prime Minister in charge of foreign trade and economics, immediately recommended it be circulated to the Politburo of the CPC and to the State Council. Deng Xiaoping, who had just re-emerged, approved it by writing "joint ventures can be set up." This is how joint ventures began in China. (China's first Joint Venture Law was introduced the following year.) This important approval led to a breakthrough in the restricted thinking of that time. It happened even before the meeting of the Third Plenary Session of the 11th Congress, which was seen as the starting point of Reform and Openness.

Interestingly, 21-year-old Deng Xiaoping had worked as a fitter in Workshop 76 of the Renault Automobile Factory during his work-study in France in 1925. He was one of the earliest first-generation CPC leaders to have contact with the automobile industry. He saw how automobiles were produced and their amazing ability to create wealth in society.

Deng Xiaoping visited Japan when he made his political comeback in 1978. When he visited the assembly line at Nissan Motor Company, he said to the owner, "I'm grateful to you for what I have learned. It has allowed me to understand what modernization is."

Also in 1978, Shanghai sent a proposal to the 1st Ministry of Machinery Industry to "import a car assembly line, reform the Shanghai Automobile Factory, and achieve annual output of 150,000 vehicles, most of which will be exported for foreign exchange." It was approved by the Ministry and by the State Council soon thereafter. Shanghai subsequently issued invitations to major auto manufacturers around the world.

It should be noted that China had tight restrictions on car consumption and there was almost no market for cars at all. The main purpose for proposing the introduction of a car assembly line was not to establish a Chinese car industry, but to export cars and earn the valuable foreign exchange urgently needed to rebuild the economy.

A delegation, led by Deputy Minister Rao Bin, visited GM on 21 March 1979 with the purpose of introducing a sedan to Shanghai. However, GM rejected the proposal. It didn't think China needed sedans nor did it have the conditions for production. The industrial base for spare parts was especially lacking.

Rao Bin then led groups to almost every major automobile company in the world. Western countries were generally cautious about the Chinese auto market. Its growth prospects were assessed as very poor. French companies felt that there were sales and foreign exchange problems. Toyota was pursuing a joint venture with a Taiwanese company. And Nissan was only willing to provide outdated models.

Volkswagen, based in Wolfsburg in Lower Saxony, was one of the very few producers interested in the Shanghai project. It was a second-tier manufacturer in the global auto industry and was hoping to find a production base in Asia to compete with Japanese cars. So, it took up the offer and began discussions. Their cooperation began with 150,000 vehicles, including commercial vehicles and Golf and Santana sedans.

However, the second global oil crisis that broke out in the early 1980s destroyed Shanghai's plans with Volkswagen. Volkswagen's board of directors suspended the project. Fortunately, Dr. Wolfram Nadebusch, Volkswagen's head of planning, and Dr. Li Wenbo argued strongly for it and the board of directors left the door open with China. However, the scale was significantly reduced to just 30,000 Golf sedans.

Rao Bin led a delegation to Volkswagen in Germany in 1979 to establish the Shanghai Sedan project. China made overtures to global partners, but only VW took up the offer. Dr. Li Wenbo of Volkswagen, second from left, later became its chief representative in China.

When CNAIC was established in 1982, Rao Bin often told me that sedans made up 70 percent of global automobile production. And China was the only car industry in the world that didn't produce them.

Production of sedans was limited until the mid-1980s. Annual output including off-road vehicles was only 5,000. This was less than the daily output of large foreign companies.

Shanghai set up a jointly managed tractor and automobile company as a subsidiary the year that CNAIC was established. Qiu Ke was Chairman of the Board and Jiang Tao was appointed preparatory team leader. The parent company supported negotiations between Shanghai and Volkswagen. After several years of talks, it became clear that Shanghai's plan to introduce technology and export cars was unrealistic.

Therefore, Rao Bin seized on the idea of setting up a joint venture. There was no precedent for it in the automobile industry. Moreover, sedans were restricted. The State Planning Commission still saw cars as non-productive and was in conflict about them. There was a heated debate with CNAIC.

The only way was to go to Deng Xiaoping. In order to avoid hitting a brick wall, CNAIC wrote a report bypassing formal channels and sent it to Deng Xiaoping through his secretary Wang Ruilin. In June 1982, Deng clearly noted his approval on the report by writing: "Joint ventures for sedans can be set up."

China then began a new round of negotiations with multinational companies. Zhang Xiaoyu, later Director of the Automotive Bureau of the Ministry of Machinery Industry, recalled his involvement in negotiations with the Vice President of GM. "I was impressed," he said. "We talked in the reception room on the first floor of CNAIC. At that time, I was just Director of the Planning Department and sat in the second row. The Chinese side proposed that the joint venture produce 20,000 to 30,000 cars annually. The Americans asked, 'Is the translation wrong? Is it a week or a month?' The translator said, 'No, it's a year.' One old gentleman shook his head, thinking we were joking. How big was the car market in China then? Less than 5,000 vehicles a year. Saying we could produce 30,000 cars was already a bold statement. In the end, the Shanghai project reported to the Planning Commission that 30,000 cars were too many. It was changed to 20,000 vehicles and 10,000 spare parts with a total investment of 318 million yuan."

1982: Volkswagen's Bridgehead in China

Similarly, in 1982, Dr. Hahn, the strategically-minded new Volkswagen chairman, was able to seize on changes in the Chinese auto industry at the right time. I interviewed him several times in Germany and Beijing. He recalled: "I became Chairman of Volkswagen in January 1982. I was confronted with the Shanghai project as soon as I started. It wasn't seen at all favorably within Volkswagen, so I decided to take charge of it myself."

"I spoke about my ideas with my colleagues," Hahn said. "Although China has just emerged from the shadow of the Cultural Revolution, it is undergoing reforms towards a market economy with Deng Xiaoping's leadership. Rapid development in all aspects will be unstoppable. Its civilization is more than 5,000 years old, and it led the world until the 18th century. After all, they are already making space launch vehicles."

Under the joint direction of Rao Bin and Hahn, the interaction between Shanghai and Volkswagen started warming up again in 1982. They decided to assemble 500 Santana sedans in Anting.

Lao Hu, leader of CNAIC's Small Vehicle Team, gave me a test ride in a dark blue Santana in the spring of 1983. It was the latest and largest model in Volkswagen's lineup at the time. It was the second-generation B-class Passat, which had only been on the market for one year. It was a best-seller in Brazil, where it was mainly produced, and had just been put into production in Spain and South Africa. Nissan was assembling them under license for the Japanese market.

Rao Bin, who made the decision to introduce the Santana, clearly understood the technical situation in the Chinese automotive industry. The Santana had a simple structure which was relatively easy to manufacture. As events proved, even with the Santana, there were huge challenges for a domestic company with a weak base like Shanghai Tractor.

The joint venture negotiations between Volkswagen and Shanghai dragged on for six years. Laws, regulations, and institutions were still being developed in the early days of Reform. The negotiation process served to perfect open systems in China.

For example, Volkswagen gave Shanghai 16 patents all at once, but after a few months had passed, the Chinese still didn't know where to apply for patent protection. This was because China didn't have such an institution at the time. Finally, Li Wenbo, who was the chief representative of Volkswagen in China, went to the German Federal Ministry of Economic Cooperation and Development and persuaded the Minister to include patent protection as a key point in cooperative projects with China. The Chinese patent protection system owed a lot to the German Patent Office.

Hard currency was needed for importing equipment, introducing technology, wages for foreign employees, and importing spare parts. China was a poor country at that time, and foreign exchange was scarce. In the 1970s, when Deng Xiaoping led the Chinese delegation to the UN General Assembly, he saved all his own US dollar spending money. There was only enough to buy some croissants when he transited via Paris. He asked the Chinese embassy to buy them for his teachers, so they could try something new. It's hard to imagine today. There was no precedent for foreign-invested joint ventures in the 1980s. The unreasonable foreign exchange balance requirements were enough to bring the project to a premature end because of the country's strict approval processes.

A decisive measure Volkswagen took was to propose the establishment of an engine plant in Shanghai with a production output three times the domestic demand. The surplus engines would be exported to Volkswagen's other factories. In this way, foreign exchange would keep flowing back. This gave the joint venture the chance to finally pass a crucial threshold for foreign exchange approval.

The fifty-fifty share ratio that continues to be used in China's auto joint ventures today was set by policy. It was also first proposed by Volkswagen in its joint venture contract with Shanghai. Shanghai Volkswagen initially had a registered capital of 255 million yuan, or about 190 million Deutsche Marks, half of which was invested by Volkswagen Germany, and the other half shared by three Chinese partners: 25 percent from Shanghai Tractor Co., 10 percent from China Automotive, and 15 percent from Bank of China Investment Company. Shanghai Automotive Industry Corporation (SAIC) has now acquired all Chinese shares.

Hahn later told me, "The fifty-fifty ratio was based on the consideration that China needs us and we also need China. China is such a vast country that you cannot understand it without visiting a few times. After all, we are outsiders and we need to cooperate and coordinate with China on a lot of things. Fifty percent each is most evenly balanced. Neither one has too much or too little. That helps both sides communicate equally. I was very pleased when the contract was signed, neither side took it out again to argue about it, and there were no ambiguities. I think that is the best kind of contract. When it's signed, it can be locked away in a cupboard."

According to the agreement, the Chairman and General Manager of Shanghai Volkswagen were to be appointed by the Chinese side, and the Deputy by Volkswagen. People in Wolfsburg complained about this, but Hahn insisted on it. In practice it proved to be a wise move.

Shanghai and Volkswagen signed the joint venture contract at the Great Hall of the People in Beijing on 10 October 1984. The groundbreaking ceremony of Shanghai Volkswagen Automotive Co., Ltd. was held in Anting two days later. German Chancellor Helmut Kohl, who was on a state visit, and Chinese Vice Premier Li Peng wielded shovels at the groundbreaking ceremony.

3. Shanghai Volkswagen: First joint venture survivor

The storm over Beijing Jeep

Beijing Jeep, Shanghai Volkswagen, and Guangzhou Peugeot, China's first-generation automobile joint ventures, were established between January 1984 and July 1985.

These three joint ventures were trial balloons started by local governments with strong experimental aims before it was decided to set up the car industry in China. Cut off from the outside world for so many years, China was poles apart from advanced world standards.

These joint ventures had almost no rulebook to follow. There were many obstacles and the chances of success were not high.

It's a shame that while Shanghai Volkswagen maintained its leading position, Beijing Jeep and Guangzhou Peugeot were no longer around.

Beijing Jeep Company stood as the earliest automobile joint venture in China.

Beijing Automobile Industry Corporation (BAIC) signed a joint venture agreement with American Motors Corporation (AMC), the fourth largest automobile manufacturer in the US, at the Great Hall of the People on 5 May 1983. The Chinese side was known for producing the Model 212 off-road vehicle. AMC was a specialist off-road vehicle company which had launched the military jeep widely used on European battlefields during World War II.

Beijing Jeep Corporation had a total investment of US$50 million. BAIC held 68.65 percent of the company in the form of plant, equipment, and a portion of the funds. AMC held 31.35 percent of the company with specialist technology, intellectual property rights, and US$16 million in cash.

The motivation behind the joint venture was the impending upgrade of the 212 off-road vehicle which equipped a lot of Chinese military units.

The Beijing 212 was the most successful independently developed military off-road vehicle in the early period. According to Chen Zutao's recollection, it was a copy of a Toyota Land Cruiser seized in the 1962 Sino-Indian border war. The military was not satisfied at first and made a lot of recommendations. Mao Zedong rode in a convertible model of the BAIC 212 twice, reviewing Red Guards from around the country on Chang'an Avenue on 18 October and 11 November 1966 at the beginning of the Cultural Revolution. After these reviews, the military informed BAIC that the vehicle design was fixed and must not be changed in any way.

Until the beginning of Reform, the 212 was not only used by the military, but was also the main vehicle of Chinese officials. Because of its unique shape, superior off-road performance, durability, and low price, there's no shortage of fans even today.

According to participants, the reason that negotiations between BAIC and AMC took so long was the difficulty in choosing a model. BAIC favored the CJ off-road vehicle then being produced by AMC. Its performance and technical specifications could improve the 212. However, the military didn't agree, saying that this vehicle was too similar to American Jeeps seen on the Korean battlefield. It said, "We had been in conflict with them for so long that we couldn't accept bringing it into production."

AMC was just bringing the Cherokee onto the market. It became the forerunner of the Sport Utility Vehicle (SUV). BAIC finally decided to introduce it because it was a comfortable 4WD vehicle with off-road capability. It was roomy inside and unlike the old Jeep in appearance. But the military never took it up because it cost a lot and it didn't meet their requirements for off-road capability.

There were still markets despite the emphasis on politics and the philosophy of struggle immediately following the Cultural Revolution. The Beijing Jeep joint venture began with a lot of conflict. Clashes occurred between Chinese personnel and management staff sent by AMC. Some veteran workers who had been in the Korean War firmly opposed the directives of US managers.

An anecdote told to me by a Chinese engineer explained why the joint venture collapsed. A US manager found a defect in a Cherokee and told the workers to scrap the vehicle. The next day he saw the vehicle was still there, so he took a sledgehammer to smash it. A group of indignant workers were about to attack him, so he had to give in. The quality of Cherokees in China still hadn't improved after 20 years.

Joint ventures were strange to the mainstream media at the time. I remember when I visited Beijing Jeep, I took a photo of the new Cherokee and published it in the *People's Daily*. Afterwards, the industry and commerce night editor told me he was criticized for it and was asked why a foreign product was being promoted in the party newspaper.

The State didn't have any preferential policies for the new joint venture companies at first. Authorities handled things according to regulations. Imported vehicle parts were subject to high tariffs of more than 100 percent. Utility departments supplying water and electricity also saw joint ventures as an opportunity not to be missed and demanded exorbitant prices. Beijing Jeep suffered serious losses and could hardly operate in that harsh environment.

Things finally came to a head in an atmosphere of mutual distrust and hostility. By November 1985, Beijing Jeep, started for just over a year, couldn't keep going. They couldn't issue letters of credit for foreign exchange because they had just gone into production and there weren't enough foreign exchange funds. 1,080 Complete Knock-down (CKD) car part sets[2] sent by AMC were held at the port and not allowed through customs, so Beijing Jeep had to stop working while they waited for them.

A few months after this stalemate, AMC Vice President Tod Clare, the original US negotiator, wrote directly to the State Council leaders to express dissatisfaction with the investment environment and ask for preferential treatment for imports and exports and on foreign exchange policies. The publication of articles by US journalists about a US auto company falling into a quagmire caused worldwide concern about the bad investment environment in China.

Clare himself came to Beijing to negotiate with the Chinese. The deadlock over Beijing Jeep embarrassed both governments, because US Treasury Secretary James A. Baker III and Chinese Vice Premier Yao Yilin were about to make reciprocal visits. Gu Mu, then Deputy Prime Minister in charge of foreign trade and economics, instructed State Economic Commission Deputy Director Zhu Rongji to deal with the matter. Chen Zutao, who had just taken over as General Manager of CNAIC; Zhang Jianmin, Director of Beijing Economic Commission; and Zhao Nailin, General Manager of Beijing Jeep, led the negotiations. Chen Zutao told me that none of the import-export and foreign exchange quota policies that Clare was asking for could be decided by just a few representatives on the Chinese side. Each had their say at the

---

[2] A complete knock-down (CKD) is used to describe a product that is sold or transported as a set of parts, which must be put together before the product can be used by the customer. Goods are shipped in CKD form to reduce freight charged on the basis of the space occupied by (volume of) the item.

negotiations, but they didn't make any progress. Clare finally announced that he was leaving and threatened to withdraw the investment. It showed how great the pressure was when Zhu Rongji called Chen Zhutao to say, "We'll both lose our jobs if the negotiations don't succeed."

Chen Zutao received a phone call from Vice Premier Gu Mu late at night. The State Council had agreed to give joint ventures preferential treatment. Chen called Zhang Jianmin and Zhao Nailin at 4 AM to hurry to the Beijing Jeep meeting room. Clare, who had gone off in a huff the day before, arrived soon afterwards. It seems no one wanted to let the cooperation break down. With the State Council's authority, Chen Zutao announced the Chinese side's commitment and the two sides finally reached an understanding. Clare, who was about to miss his flight, signed his name on a blank sheet of paper and said, "Mr. Chen, I'm trusting you. You write the agreement."

Chen Zutao never revealed to me what the Chinese side promised, but I learned through other channels that they gave the US side some foreign exchange compensation, and in the following years they gave Beijing Jeep a foreign exchange quota of US$17 million a year. Although the storm had subsided, the confidence of the US side had suffered damage. For many years after that, they held onto the view that they would neither withdraw nor increase their investment.

The Cherokee is a high-end recreational vehicle for the American middle class. It's expensive, consumes a lot of fuel, and isn't as practical as a sedan. Beijing Jeep was destined from its inception to struggle in an overly narrow market sector.

AMC merged with Chrysler in 1985. In 1998, Chrysler and Daimler were reorganized as the DaimlerChrysler Group. The frequently changing new owners always saw Beijing Jeep as of little value, until they finally gave it up.

Martin Posth and Volkswagen's 1,000 days in Shanghai

Martin Posth, a member of Audi's Executive Management Board, was sent to Shanghai with Hahn's direct intervention. He was partnered with Hans Bauer, a department head at Volkswagen's Kassel factory.

Shanghai Volkswagen could only assemble two Santanas a day at that time. In Germany, a single Volkswagen assembly shop could produce at least 1,000 cars a day. The well-known Workshop 54 in Wolfsburg could produce 3,000 vehicles per day during its heyday. The goal given to Posth and Bauer by Hahn in the name of the Volkswagen Supervisory Board was clear and simple, "Start everything again from scratch and build Workshop 55!" Perhaps not more than ten people at Wolfsburg believed then that his vision could become reality.

Posth later recalled in a book entitled *1000 Days in Shanghai* that in a city with a population of 11 million then, there were only three public petrol stations. Santana's assembly workshop was in two old museum-class factory buildings on the verge of collapse that Shanghai Trailer Co., Ltd. had contributed as investment in kind, and the assembly of the old Shanghai sedan was crowded into these buildings along with the Santana.

"I stopped breathing and just stared at the dilapidated factory buildings. My mind went blank. Are these dusty shacks really an automobile factory? And is this where VW wants to make cars with the Chinese?" Posth later recalled.

The first Executive Committee of Shanghai Volkswagen was confirmed by the first Board of Directors as: Zhang Changmou, General Manager; Posth, Deputy General Manager and Executive Business Manager; Bauer, Executive Technical Manager; and Fei Chenrong, Personnel Manager.

When I paid a visit to Shanghai Volkswagen for the first time in 1986, there was no production line in the factory. A white-bodied Santana was held by a hoist and was pushed on a welded tubular steel trolley from one station to the next as it was being assembled. The general manager's office was located in a crumbling old building.

There was no air conditioning in the offices of the senior management of the joint venture. They were drenched in sweat in summer and chilled to the bone in winter. Electricity was in short supply and either the heating stove in the office could be turned on or the paint shop baking oven could be used. Only one or the other could be operated at a time. The biggest luxury was a refrigerator in the summer and a kettle for making hot tea in winter. There was only one telephone for the entire company. All the office furniture was made from boards taken from the crates the automobile parts came in.

Unforeseen problems awaited Posth every morning. Urgently needed equipment which had been in port for 10 weeks couldn't be transported out because of customs problems; workers discharged toxic fluids directly into the river; workers hung their washing to dry between the buildings; water outages were caused by farmers in the surrounding area who tapped into the water supply for their vegetable fields; and dust raised by people, cars, and buffaloes using the road which ran through the factory area affected the quality of the paint...

Rao Bin and Dr. Carl Hahn signed the Shanghai Volkswagen joint venture agreement at the Great Hall of the People in Beijing in 1984.

Posth and the Chinese began to set up a corporate system. They introduced VW's management methods for mass assembly in Japan. They finally got production of the Shanghai away from the factory by simply building a new factory. And on the morning after the Mayor's first site visit, a giant crane was used to block the chaotic passage of people and vehicles through the factory area.

Posth had a strongly held belief that "If we want to produce products of the same quality all over the world, then we should have personnel of the same quality." Germany attached the highest priority to senior technicians, but professional training was clearly ignored in the negotiations. The Chinese liked to guarantee that the factory had all the best employees, but they had never done any analysis of that.

So Posth returned to Wolfsburg and appealed strongly for the establishment of a German-style dual-track mechanic training program for Shanghai Volkswagen. The director of the training center at Wolfsburg estimated that standard (basic) training would require an investment of 7.5 million Deutsche Marks. The joint venture was strapped for cash and the money was not available. Fortunately, the Technical Cooperation Agency (GTZ) of the German Federal Ministry of Economic Cooperation was willing to provide assistance. A two-track vocational training center was established in Shanghai with a training director sent from Wolfsburg. This center for 250 young people a year was later used by the Chinese government as a model for training skilled workers.

As well as training technicians, the joint venture hoped to give other Chinese employees the same opportunities. From the mid-1980s, a large number of Chinese management and technical personnel went to the headquarters in Wolfsburg, as well as to other Volkswagen locations, for training and study. Hahn once recalled that there were no fewer than 100 Chinese people at Wolfsburg at that time. Among other things, they became familiar with modern production methods and international accounting rules.

Posth found that German employees from Wolfsburg didn't see themselves as members of the Shanghai Volkswagen team. The underlying sense of "we are the greatest because we are from VW" was deeply ingrained. Shanghai Volkswagen's management decisions were often discounted. They argued, "My boss in Wolfsburg says..." Posth and Bauer immediately interrupted them by saying, "Your bosses are all here in Shanghai. We have to understand that we're working for Shanghai Volkswagen. If not, we won't be able to complete the mission."

Although by German standards Shanghai Volkswagen was not flourishing and outsiders were doubtful about its future, Posth said, "All our employees, Chinese or German, have the same dream — to build China's largest, most advanced, and best car manufacturer. The only thing left to do is to make our common dream come true."

China had been cut off from the rest of the world for 30 years. No one had foreseen the misunderstandings and conflicts caused by cultural differences in this unprecedented joint venture. Ms. Chen Yunqiu, who served as a translator at Shanghai Volkswagen, later recalled, "Germans are straightforward and matter of fact, but Chinese talk in a roundabout way." Bauer wanted a yes or no answer on whether or not a job was completed, but his Chinese counterpart would reply, "We are doing our best to coordinate it." After several rounds of questions and answers, Bauer couldn't help himself but asked, "Ms. Chen, is your translation correct?" She asked her Chinese compatriots to give clear answers and the Chinese replied sharply that she was exceeding her authority.

Differences in values made communications between the two sides difficult. At one technical negotiation, VW headquarters asked the Chinese to pay a large fee for consultation services. The Chinese flew into a rage. They felt that consultations shouldn't cost anything. Even Ms. Chen couldn't be impartial and argued directly with the Germans in German. "Bauer was so annoyed that he clenched his fists. I regret that now," Chen told Posth many years later.

Posth said in his memoir, "80 percent of the problems which lead to the failure of joint ventures don't come from the product or the cost accounting. These companies don't lose to competition or to the market. They lose because people can't communicate. It's still a big problem."

High standards or being held in check

Many would consider it a great thing to be general manager of a Chinese automobile joint venture, but that was far from the truth. Almost all the Chinese general managers I was in contact with in those years struggled along, stigmatized as "foreign compradors." The pressure was more than a person could bear. The suicide of Fang Hong, third General Manager of Shanghai Volkswagen, and the departure of Song Zuwei, first General Manager of Shenlong Company, could only be comprehended by understanding the pressure on each of them. They were healthy, strong, and open-minded intellectuals.

The person I had the most contact with was Wang Rongjun, the second General Manager of Shanghai Volkswagen. He had been involved with setting up FAW and SAW. He was a talented man recommended by Chen Zutao, President of the CNAF, at the behest of Shanghai Mayor Jiang Zemin.

The biggest pressure Wang encountered after he took over was the slow progress in localizing Santana cars. According to the contract, locally sourced parts for the Santana had to be sent to Wolfsburg for technical certification. The first batch of Shanghai Volkswagen's parts and components were mostly from small, back alley factories in Shanghai, and it was by no means easy to reach VW's standards. CKD (component assembly) had been going on for more than two years and the localization rate was only 2.7 percent. Only wheels, radio-cassette recorders, and antennas for the Santana were made in China.

When the media reported that localization rates at several other joint ventures were surging while Santana was still working at it, VW headquarters in Wolfsburg was beside itself, not to mention Wang Rongjun.

Certification of parts by VW was certainly a mess. Twenty years ago, the instinctive response was that the Germans intended to choke us and force us to buy their fully assembled vehicles forever. Chinese managers were labeled foreign compradors dependent on the foreign side, and most seriously were accused of humiliating China by surrendering its sovereignty.

Apart from these criticisms, one consideration was that German Volkswagen's standards were too high for most Chinese parts factories. Yet there could be flexibility since Santanas were mainly sold on the domestic market. Their quality could be advanced in stages and standards didn't need to be set so high.

Shanghai Volkswagen shared a simple assembly shop with the old "Shanghai" in the early days.

Santana's first assembly line.

But the Germans were stubborn, acting strictly in accordance with the contract. They said, "Cars are the most expensive industrial product. Consumers who have spent their hard-earned money should get a Santana that performs safely, with quality and product appearance in line with international standards."

This angered some people, who said, "Let's not use their Santana brand."

Rao Bin and Meng Shaonong, the only member of the Chinese Academy of Sciences from the automobile industry, said to me more than once, "We are still beginners at making sedans and we will have to learn as we go along. The original intention of our joint venture is to achieve world standards."

Wang Rongjun said more plainly, "Why bother introducing the Santana if you don't internationalize local standards? Our old Shanghai model was made entirely in this country. Shanghai Volkswagen is an independent company, not a subsidiary of German Volkswagen. The State only approved the assembly of 89,000 Santanas from imported parts. If the parts have all been assembled and domestic parts still haven't arrived, the Germans will have to pack up and leave. They are as anxious as we are about the localization issue."

Zhu Rongji, then Deputy Director of the State Economic Commission, made a special trip to Shanghai in June 1986. Before he arrived, he heard that the Chinese general manager of Shanghai Volkswagen had helped the Germans block the parts factory, putting a drag on the progress of localization. When he heard this report at the Jin Jiang Hotel, he said coldly to Wang Rongjun: "I've heard about you. You've been doing well for yourself."

Wang Rongjun could have become a "yes man" on hearing this, but he argued back, "Our parts and components factories have outdated equipment, backward technology, and a gap of at least 30 years with Germany. For every Santana component to meet German standards, nearly all of the equipment and technology would have to be brought in. Doing it properly will take time and money."

A few days later, he showed Zhu Rongji a photo of a tire on a test machine. After running at high speed, splits appeared in the rubber and the tire cords burst open. Wang Rongjun said, "The tires were certified by VW of Germany, but they still have serious quality problems. The difficulties of localization cannot be underestimated."

Zhu Rongji was lost in thought as he held the photo. The field inspection over the past few days had given him a new understanding of the arduousness of the task. He said, "It is like not knowing when a volcano will erupt. You can't imagine the consequences."

At the end of the trip, Zhu Rongji came to a new decision about the complaints. He said, "In localizing Santana parts, we have to conform 100 percent to Volkswagen Germany's standards. We can't lower them even by 0.1 percent."

Shanghai Volkswagen got the Chinese car industry off to a good start by keeping high standards. Its greatest contribution was to withstand the pressures of domestic ignorance and extremism, creating high standards for China's cars, its auto parts system, and its ability to keep pace with the rest of the world. If standards had been allowed to float along at the beginning, the Chinese car industry would probably be only third or fourth rate today!

The three armies march on, each face glowing[3]

There was high level approval for what Shanghai Volkswagen was doing.

A Shanghai Santana Car Localization Group was established across regions and sectors at the initiative of Zhu Rongji. He took over as the mayor of Shanghai early in 1988 and became a staunch supporter of localization. He said, "The number of Chinese products which have really reached international standards can be counted on your fingers. The Santana is one of them and is the hope of Shanghai."

Jiang Zemin, then secretary of the Shanghai Municipal Committee of the CPC, proposed that the Santana Group should not just benefit Shanghai, but must go nationwide.

---

[3] From a poem by Mao Zedong called *The Long March*. This poem was written toward the end of 1935 when the Long March was almost finished. The last line of the poem (used here) refers to the feeling of accomplishment by soldiers at having overcome obstacles and achieving victory.

Hahn still praises the policy-making departments for organizing the participation by Third Front aerospace companies[4] with strong technical prowess in the localization of Santana components.

The existing domestic system was committed to investing in original equipment manufacturers (OEMs).[5] Components factories needed to carry out technological transformation, but the shortage of funds became an insoluble problem. However, the Shanghai people were shrewd, and beginning in 1988, with the approval of the State Council, the price for each Santana was increased by 28,000 yuan, for what was called the localization fund. Even the old Shanghai sedan which had no imported components contributed 5,000 yuan to the fund for each vehicle sold. Fifteen thousand Santanas were produced a year and the fund reached 400 million yuan.

The State extended this measure to several other cars in 1990, adjusting the amount to 23,000 yuan per vehicle. By 1994, when the State suspended the fund, the Shanghai Santana had already raised 4 billion yuan. This massive amount turned the situation around and allowed component companies to bring in technology and equipment.

Reaching Volkswagen Germany's standards was a process of change for Shanghai Volkswagen. When the directors and engineers of the parts factories recalled the impact of VW standards in those years, they said that their hardships were like a famous saying in Aleksey Tolstoy's[6]

---

[4] The Third Front Movement (三线建设; Sānxiàn jiànshè) was a massive industrial development by China in its interior starting in 1964. It involved large-scale investment in national defense, technology, basic industries including manufacturing, mining, metal, and electricity, transportation, and other infrastructure investments. It was motivated by national defense considerations, most noticeably the escalation of the Vietnam War after the Gulf of Tonkin Incident, the Sino-Soviet Split, and small-scale armed skirmishes between the two countries.

[5] OEM is an abbreviation for original equipment manufacturer. In the automotive industry, this term generally refers to automotive manufacturers. It can also refer to the manufacturer of origin for any product.

[6] Aleksey Nikolayevich Tolstoy (b. 29 December 1882–d. 23 February 1945), nicknamed the Comrade Count, was a Russian and Soviet writer who wrote in many genres but specialized in science fiction and in historical novels. He was the son of

novel *The Ordeal* — "soak three times in salt water, bathe three times in blood, and cook three times in potash."

There were a lot of problems raising the Santana's localization rate by even one percent. All the manufacturers were worried. In the past, there were only six test criteria for steering wheels, but there were more than a hundred for the Santana. "We've never seen such high standards!" they said.

The prototypes were returned again and again at the beginning. All the companies, without exception, cursed the Germans for being strict and inflexible and deliberately making things difficult. There were many conflicts and arguments in the board of Shanghai Volkswagen, in the Quality Assurance Department, and at the production site.

Hundreds of components factories in Shanghai, Beijing, Nanjing, Hubei, Jilin, and Guizhou faced failures and returns, and experimented day and night to solve the problems. Some factories also set up workshops in special zones, where workers could only go on duty after strict training.

Volkswagen organized some retired experts to go to China to bring the components factories up to German standards. Wang Rongzhen recalled that they were unpaid, and hundreds of them came to help solve a lot of technical and managerial problems. They played a big role in speeding up localization.

For Volkswagen Germany to certify a part, it needed to go through an 18-step process, from inspection of the plant's equipment to the acceptance of the item. Three test samples had to be sent. The first was called the initial sample. After it passed, an off-tool sample[7] was sent. And finally, a batch production sample was produced on the production line. It's thorough and detailed.

Parts factories, which had gone through this painful effort and were finally approved by Volkswagen, took great pride in their achievement.

Count Nikolay Alexandrovich Tolstoy (1849–1900) and Alexandra Leontievna Turgeneva (1854–1906).

[7] Off-tool sample (OTS) is an initial sample created using production tooling. It is used to check design and fine tune tooling prior to making production quantities.

In the past, the parts factories promoted their products while relying on the reputation of the OEM. When they were approved by Volkswagen, other automakers came to place orders, and without inspection.

Shanghai Santana's annual output exceeded 100,000 vehicles in the early 1990s. When the localization rate reached 90 percent, Shanghai Volkswagen and its parts and components localization complex were glad to have overcome all the obstacles.

The introduction and localization of Santana drove rapid improvement in the overall standards of China's automobile industry. An obvious parallel would be that of Mexico's production of the Volkswagen Beetle. Over the past 30 years, the localization rate had only been 60 percent. Shanghai took just over a decade to reach 90 percent with the Santana.

Deng Xiaoping came to Shanghai Volkswagen and saw a shining new Santana drive from a computer-controlled production line in the spring of 1991. He said emotionally, "If it wasn't for Reform, we would still be beating them out with hammers. It's all different now. This is a qualitative change!"

Deng Xiaoping inspecting Shanghai Volkswagen on 6 February 1991. He said, "If not for Reform and Openness, we'd still be beating out cars with hammers."

01 I want to be a bridge: An impression of Rao Bin

A last conversation

It was unbearably hot at the height of summer in July 1987. Rao Bin was over 70 years old and had already left his leadership position in the Chinese automobile industry. Compelled by a sense of failure, he hurried to Shanghai to investigate several component factories.

Shortly after returning from the forum in Shiyan, he heard there had been reports to Beijing that Shanghai Volkswagen's German partner was being difficult about quality problems with Santana parts. Rao Bin, an old man who had been in the car industry all his life, stayed calm about it. He said that producing the Santana was like an entrance examination for the establishment of a modern car industry in China — you had to do the hard work and get the quality first to enjoy the rewards later.

He was the one who orchestrated the cooperation with German Volkswagen to produce the Santana sedan. His intention was to bring forward high standards which were world recognized, not decided on behind closed doors by Chinese people.

On the evening of July 30, Shanghai Mayor Jiang Zemin hosted Rao Bin at the Hengshan Hotel. In the 1950s, Rao had been the director of FAW and was Jiang Zemin's former superior. That afternoon, Rao shut himself in his hotel room and drafted his talking points. Although he was retired, he didn't see the meeting as just a chance to talk about old times. According to people who later saw the outline, it was a comprehensive vision of how to develop the car industry in Shanghai. That night, Jiang Zemin accompanied Rao Bin back to his hotel room, and they talked about cars and their own career difficulties until very late. Finally, he walked Mayor Jiang to the elevator.

Rao Bin didn't get up again the next day. Jiang Zemin was the last person to speak with him.

When I compiled a biography of Rao Bin for broadcast by Xinhua News Agency, I called him an outstanding founder and pioneer of China's automobile industry.

Foreigners refer to him as "the father of Chinese cars."

The years of tempering

At Mengjiatun in Changchun on 15 July 1953, on the waste ground of a Japanese bacterial brigade where the Kanto Army had murdered anti-Japanese idealists, the valiant Rao Bin threw the first shovelful of black soil onto the foundation stone. Two lines of red characters were engraved on the stone: Foundation stone of the First Auto Works, Dedicated to Mao Zedong.

Standing behind Rao Bin were Soviet factory design experts: Hu Liang, who had chosen the location; and Li Gang and Chen Zutao, members of a purchasing team stationed in the Soviet Union. US-educated Meng Shaonong, the first automobile expert to join the project, couldn't make it because his train was held up by a flood.

That was the purest, most selfless, and most effective era in the history of New China's construction. Ten large factory buildings were constructed and equipment were installed in severe winter cold of minus 30 degrees Celsius. Amidst the sounds of singing, laughter, and volunteer labor slogans, there was the tall figure of the young factory director. He wasn't yet 40 years old, but had been Mayor of Harbin and Deputy Secretary of the Songjiang Provincial Party Committee.

China paid a lot of attention to FAW. Thousands of cadres were sent from all over the country. Promising young county leaders became technicians there. It was a technical academy, and Rao Bin was its principal and also a student. His slogan was "Don't be unprofessional." He had studied medicine at university but he employed experts to be the teachers. He pored over a dozen business management and automotive engineering courses, working late every night. Between construction and study, the FAW plant was a city that never slept. Thirty years later, that tempering experience was in the career resumes of many factory directors, governors, ministers, and Central Committee members.

It can't be denied that FAW was made by introducing a complete package of Soviet technology, equipment, and products. The Soviet Union copied its ace product, the Moscow Likhachov Automobile Factory's ZIS[8] brand four-ton truck in Changchun. The first Liberation CA10 four-ton truck drove off the production line on 15 July 1957. It was followed by a second and a third, until the 1 millionth 30 years later. Rao Bin stood with Mao Zedong, Zhou Enlai, Zhu De, and Deng Xiaoping at the end of the production line, smiling with satisfaction.

In 1958, the slogan "One day is equal to twenty years" set off the enthusiasm of the Great Leap Forward, and "Catch up with Great Britain and the United States" immediately became the consensus disseminated in newspapers and magazines. The US was the kingdom of cars. So, in March, Rao Bin received a task directly from the Central Committee of the Communist Party to produce small cars immediately.

Rao Bin led the people at FAW to create a miracle. The Dongfeng sedan was born just 23 days after the mobilization meeting. In mid-May, the leaders at the second meeting of the Eighth Party Congress in

---

[8] The ZIS-150 was a Soviet truck. In 1947 it replaced the ZIS-5 truck on the assembly line. ZIS-150 together with GAZ-51 was the main Soviet truck of the 1950s, judging by their quantity. ZIS-150 was also manufactured in China as the Jie Fang CA-10 at First Automobile Works.

Beijing proposed that they should have a look at the first car in New China. Rao Bin personally drove the Dongfeng into Zhongnanhai. When the participants walked out of Huairen Hall during a break in the meeting, Mao Zedong took the aged Lin Boqu with him for a ride in the car to the applause of the delegates.

Rao Bin hurried back to FAW straight away, because he had been given an even more important task — to immediately trial manufacture the advanced Red Flag sedan for the Central leadership to travel in. Mobilizing the entire factory staff, he hung all the drawings in the auditorium and called on the best and brightest people to take them and make prototypes. The 3,400 drawings were all taken by skilled craftsmen within a few days and a month later the Red Flag 770 with its lantern-shaped taillights was launched.

Deng Xiaoping and Li Fuchun inspected FAW shortly afterwards. Deng Xiaoping asked Rao Bin, "What's the Red Flag like compared to the Volga?"

"It's of a higher standard," Rao Bin replied.

"And compared to the ZIM?"[9] The ZIM was a luxury sedan made in the Soviet Union especially for senior cadres.

"It's also of a higher standard."

Hearing Rao's answers, Deng Xiaoping said humorously, "Well, if it's better than the ZIM, you can produce more of them. They can burn alcohol if there's not enough gasoline. There are sweet potatoes to make alcohol from. Just make sure you don't burn Maotai."

In 1964, after three years of famine, Mao Zedong said it was time to build a second automobile factory. And so, Rao Bin was transferred to Shiyan among the mountain ridges of Shennongjia[10] in November that year to start over a second time.

Rao Bin had three policies which summed up FAW's experience with the previous car: SAW could not just make one model but must

---

[9] The ZIM-12 (Russian: ЗиМ-12) was a Soviet limousine produced by the Gorky Automotive Plant from 1950 till 1960.

[10] Shennongjia Forestry District is a county-level district in northwest Hubei Province.

serialize models from one to eight ton; it had to have the latest equipment by choosing the best from machine factories throughout the country; and specialized factories would be built the same way as SAW's new factory by contracting an inland factory to do it.

However, Rao Bin was accused of pursuing large-scale projects and foreign technology during the Cultural Revolution which followed. Ignorance and fervor annihilated human nature to an unprecedented degree. He was forced to kneel and was beaten. A large wooden sign weighing more than 20 kilograms was hung around his neck with thin wire. All around him was a red sea of Mao Zedong's quotations. He rose and sank in the revolutionary flood for a full 10 years, sometimes the target of attacks and at other times dragged out to work.

It wasn't until October 1977, one year after the overthrow of the Gang of Four, that Rao Bin officially became Factory Director and finalized the design of the Dongfeng five-ton civilian truck.

Dreaming about cars

Rao Bin took up the post of Minister in the 1st Ministry of Machinery Industry in the early 1980s. I was assigned to cover the machinery industry by Xinhua News Agency. There wasn't much media at the time and there were even fewer journalists covering the economy. I spent more time on interviews than in the office. Minister Rao Bin was open-minded, had sharp views, and treated people kindly. We became friends despite our age difference.

He became chairman of the newly formed CNAIC on 8 May 1982. The auto market finally began to develop positively in the 1980s under his leadership. Rao Bin, who had worked on trucks most of his life, told me that without cars the automobile industry would only limp along. At the company's beginning, he firmly supported Shanghai in bringing in technology, transforming the old Shanghai car, and establishing a modern car industry. Faced with strong resistance from departmental authorities, he got the idea of handing a car production report directly to Deng Xiaoping, so that he could put forward the idea of producing cars through the establishment of joint ventures. In June 1982, Deng

Xiaoping clearly noted on the report: "We can have joint ventures for cars." I happened to go to Rao Bin's south facing office on the third floor of the CNAIC on the day he got Deng Xiaoping's instructions. There was a smile on the old man's face that was not seen for a long time.

Two joint ventures, Beijing Jeep and Shanghai Volkswagen, were satisfactorily established after being pushed through by Rao Bin.

The biggest demand in the Chinese market in the 1980s was for official cars. They considered offering the Audi 100 to the general public, but it was Rao Bin who advocated choosing the Santana. He argued everywhere that the model really needed in China at that time was not a luxury car like the Mercedes-Benz, but a car that was fuel-efficient, inexpensive, and safe. The Santana, a four-door sedan, was simple in structure and of moderate size, making it suitable for a wide range of business vehicles and taxis. Even today, I don't agree that to be characteristically Chinese, the imported models have to be lengthened. This was probably influenced by Rao Bin at the time.

Dr. Carl Hahn, then chairman of Volkswagen, later spoke of his negotiating opponent, Rao Bin: "He is a talented engineer, a smart manager, and a visionary strategist. Despite the differences in our origins and experience, we quickly found a human connection."

Rao Bin attended a car development seminar held at the Checheng Hotel in Shiyan in the summer of 1987. There were differences of opinion at the meeting about what form the car industry should take. I asked Rao Bin, who had already retired a second time, to give an address and talk about his views. Later, I apologized for changing my appointment. I had to leave early because something came up in Beijing. He smiled and said, "It doesn't matter, you'll have plenty to write about. Cars will be a very big chapter in China's auto industry."

In the intense summer heat, Rao Bin went to Shanghai to investigate the Santana's localization process despite suffering from cardiovascular and cerebrovascular disease. He was hospitalized and in a coma for a month. He died on 29 August, aged 74. The goodbyes we said at Shiyan were our last.

At the interment ceremony in Beijing, FAW comrades told me that before going to Shanghai, Rao Bin attended a celebration for the Jiefang truck model change. At the event, he suddenly started talking excitedly about cars, "I am old, and I can't take part in the third stage of the Chinese automobile industry. However, I'm willing to lie down and be a bridge for everyone to walk across. Let's work together to make cars and fulfill the dream we have had for generations."

Everyone in the hall fell silent. They saw tears shining in his eyes. He was crying.

Rao Bin left us and took along a whole era with him.

Chapter 2

The Car Industry Gets Its Birth Permit

I postponed some interviews and took the train to Shiyan in Hubei Province in May 1987. In the sleeper compartment, I met up with an automobile industry economist, a departmental authority, and a car factory manager. They were all enthusiastic about automobiles. One joked that it was a special train for automobile people. Key figures from the Chinese automobile industry seemed to be gathering at Wudang Mountain.

Wudang is a sacred Taoist mountain where people escape the hustle and bustle of the world. It is in central Hubei, a thoroughfare for 9 provinces. China's SAW sprang up at the foot of the mountain. The rumble of our wheels broke thousands of years of silence in the steep mountains.

I had been in countless meetings as an economics journalist, but none gave me such a strong sense of participation as this one. I had just published a special feature, "Should China Develop a Car?" in the *Outlook* magazine. It was the first time a positive conclusion had been reached on this issue in a government publication. I was the only Beijing reporter invited to this off-the-record meeting.

To avoid giving any offense, the meeting held at the Car City Hotel in Shiyan was given the rather innocuous name of China Automotive Strategy Seminar. However, the experts had a sense of mission and only one thought — cars.

After the event, the conference was called the Sedan Feasibility Seminar. It destroyed the myth that China couldn't develop a car and became a watershed in the history of China's automobile industry.

1. Summer 1987: The thaw

New China refuses sedans

Although the first domestic single-engine biplane was produced at the Beiyang government's Beijing Nanyuan Aviation School in 1914; a 10,000-ton armored steel ship was built in Fujian's Mawei Shipyard in 1912; and some reports said that Zhang Xueliang had assembled a car in the Northeast, the Chinese did not actually produce a car until the 1950s.

Shanghai used to be called the International Car Show because it had the largest car ownership in the Far East. Ford, Chevrolet, Buick, and Dodge were well-known in the busy metropolis. Ford had plans to build a car in China before the outbreak of World War II.

The Chinese who had stood up when the People's Republic of China was founded in 1949 were determined to part ways with the car. This was particularly so following the outbreak of the War to Resist the US and Aid Korea (the Korean War 1950–1953) which made US imperialists a deadly enemy. The sedan was a symbol of the US, so it had to be spurned by Chinese people. The import market for cars from capitalist countries died a natural death in the 1950s. National capitalists had to rely on spare parts hand-made by craftsmen in Shanghai to maintain the last of their imported cars.

The ZIS and ZIM[1] automobiles provided by Stalin solved the problem for leaders and ministers. There was no plan for producing sedans in the first five-year plan. Cars were allocated to officials strictly according to their rank, and the office vehicle at county level was really just a

---

[1] ZIS-110 was a limousine from the USSR's ZIL factory introduced in 1946. The 110 was developed by reverse engineering a 1942 Packard Super Eight in 1944. These cars were often given away as gifts to foreign communist leaders such as Chinese leader Mao Zedong and North Korean premier Kim Il-sung. Before production began of the Hongqi (Red Flag), Chinese leaders used various foreign luxury cars. For example, Chairman Mao Zedong used a bullet-proofed ZIS-115 donated by Stalin, while Premier Zhou Enlai used a GAZ-12 ZIM, later a ZIL-115 and Foreign Minister Chen Yi used a Mercedes-Benz 600 Pullman before Hongqi began production of its CA770.

bicycle at the beginning. It was considered really good if a county had one or two old American-made jeeps captured on Korean battlefields.

However, the next page of history was incredible because it was an unexpected derailment of China's fixed policy of refusing to have cars. It was a sudden shock.

Mao Zedong led a Party and Government delegation to Moscow to attend the 40th anniversary of the October Revolution in November 1957. This was the second time in his life that he had gone abroad. The leaders of 71 Communist and Workers' Parties from around the world gathered in Lenin's hometown to hear Khrushchev's amazing claim that the Soviet Union would catch up with the US, its Cold War rival, within 15 years. Then in cold, snowy Moscow, Mao Zedong set an aggressive goal — China would catch up with the UK in terms of steel and other major products in 15 years.

The following year, the slogan "One day equals twenty years" made the Great Leap Forward even livelier. "Overtake the UK to catch up with the US" became the most conspicuous slogan in newspapers and periodicals, and even on the walls of Beijing's streets and lanes.

So FAW had a campaign with its Dongfeng (East Wind) and Red Flag sedans to catch up with the US, a kingdom of cars.

Just over a month after the Dongfeng was driven into Zhongnanhai[2], the newspapers reported that two more cars, the Jinggangshan[3] and the Heping (Peace),[4] were born in Beijing and Tianjin respectively.

---

[2] Zhongnanhai (Chinese: 中南海; pinyin: Zhōngnánhǎi; literally: 'Middle and Southern Seas') is a former imperial garden in the Imperial City, Beijing, adjacent to the Forbidden City. It serves as the central headquarters for the Communist Party of China and the State Council (Central Government) of China. Zhongnanhai houses the office of the General Secretary of the Communist Party of China (paramount leader) and Premier of the People's Republic of China.

[3] Jinggangshan cars (based on the Volkswagen Beetle) were built by Beijing Auto Works (BAW). The Jinggang Mountains after which the car was named is known as the birthplace of the Chinese Red Army (predecessor of the People's Liberation Army) and the "cradle of the Chinese revolution."

[4] Heping (Peace) automobiles were built in Tianjin beginning in 1958 and were based on the Toyota Toyopet Crown RS-series from Japan.

Of course, only a few prototype cars could be produced, but they filled people with enthusiasm. Meetings were held all around the country to criticize conservatism and to celebrate the 40 odd indigenous cars that poured forth. For example, there was an agricultural vehicle, which could use gasoline, distilled liquor, coal, or firewood as fuel. When a repair factory built a car, they used a homemade steel furnace to refine their own iron and steel.

A car after all is a product of tens of thousands of parts and embodies the technical standards of many related industries. But enthusiasm relies on material support. The recession which followed the Great Leap Forward turned enthusiasm for cars into thin air, and cars soon disappeared. China's policy of not making sedans was back on track.

A conspicuous black hole of demand

Maximum annual production of small Chinese cars — the Red Flag, the Shanghai and the Beijing 212 off-road vehicle — did not exceed 5,000 until the early 1980s.

China's economy recovered rapidly after Reform began. When the rule that county-level officials and below could only use jeeps was removed in 1984, it was like opening the floodgates and the number of vehicles used for official purposes suddenly shot up. Officials competed to buy cars, and it soon got out of hand. Leadership in poor counties misappropriated relief and education funds to buy cars. Directors of loss-making enterprises could not pay their workers' wages, but used loans to buy luxury sedans. Such disclosures were frequently found in the news. The demand for cars set off a wave of imports and hundreds of thousands of foreign cars poured in through legal and illegal channels, filling the huge vacuum created by the lack of a car industry.

The tide of imports reached its peak in 1985, dominated by Japanese cars. 345,000 vehicles were imported that year. Factories reassembling dismantled, smuggled vehicles were everywhere in the Pearl River Delta. Hainan Island was crowded with smuggled cars. When the Central Government investigated the ban on smuggling in Hainan at the end of the year, there were 60,000 cars on the Haikou wharf waiting to be loaded for shipment to the mainland.

The pride of the Chinese people was hurt. There was no money to develop the automobile industry. However, the foreign exchange that flowed out was more than two times the total invested in the automobile industry over 30 years.

A State Council vice premier was upset about it and proposed to slash import approvals immediately. However, it was difficult to stem the tide of imported cars. Applications to import them poured into departmental authorities.

It was just at that time that prices were relaxed and inflation picked up. There were complaints about profiteering officials, corruption, shortages, rising prices, and people seen going about in shiny imported cars. Needless to say, they were condemned verbally and in writing. College students in some cities made a fuss about it. Confiscation of illicitly imported Japanese cars became the theme of the day.

In the mid-1980s, the ever-increasing enthusiasm for imported cars became a topic of anger and criticism among delegates and committee members at the National People's Congress (NPC) and the Chinese People's Political Consultative Conference (CPPCC) held in Beijing every spring. Observant journalists counted the number of cars in front of the Great Hall of the People at the two meetings in 1988. There was a total of 556 vehicles, including 495 imported cars, 24 from Sino-foreign joint ventures, and only 37 domestic cars. At the end of the CPPCC meeting, a member named Wang Zhou stood up bravely and declared, "I suggest inserting a note in the meeting documents encouraging leading cadres to ride in domestic, economy sedans."

To build cars or to buy them had become a political issue involving national pride that raised concern throughout society, and decision-makers began to pay attention.

The conference of key automobile figures at Shiyan

The China Automotive Strategy Seminar was held in Shiyan in Hubei in May 1987 under the authority of the State Council. The purpose of the meeting was to decide whether China should establish a car industry. The chairperson was Ma Hong, an economist who had just left the post of President of the Chinese Academy of Social Sciences (CASS) and was convener of the State Council Policy Advisory Coordination Group. Representatives of the State Economic Commission, the Planning Commission, the Science and Technology Commission, economic authorities, veterans of the automobile industry, as well as key automobile factories, raw materials, machinery, electronics, petroleum, transportation, municipal government, and other upstream and downstream departments each gave a read-out of their research results. Even two auto companies, Toyota and Nissan, were invited to send experts. They made a special trip from Japan but avoided giving advice at the conference.

Although people had different ways of thinking and argued their points emotionally, their conclusions were surprisingly consistent in recommending the immediate establishment of a domestic car industry.

Zhang Wei, of the State Council's Economic, Technological and Social Development Center, managed the summary documents submitted to the State Council. He collected recommendations from experts who said that a comprehensive, high quality automobile industry was out of the question without producing sedans. "We are already faced with the serious problem of whether to take hold of the huge market for sedans or hand it over to foreigners. There should be a gradual shift to sedan and parts industries from now on."

The document said that the ratio of cars to trucks should be 4:6 by the year 2000. According to forecasts, the annual demand for domestic cars in 2000 ranged from planned production of 2 million (high), to 1 million (medium), or 600,000–800,000 (low). Most experts at the meeting preferred the low figure. That is, by the year 2000, annual output of vehicles would be 1.7 million to 2.2 million. And annual output of sedans would be 700,000. Taking into account the country's financial and material resources, and construction difficulties, it still wouldn't be easy to meet the low number. (Author's note: The actual total vehicle production in 2000 was 2 million, including 600,000 sedans.)

The evidence showed that it would still be possible with hard work, although there was a gap between current technical standards and production capacity, with constraints on fuel, raw materials, roads, urban infrastructure, and the machinery and other related industries. But the car industry would support these other industries.

However, in terms of vehicle consumption, which was subject to ideological constraints, the mainstream official view was that the automotive industry should cater to official cars and taxis. Private cars would not be considered until 2010.

Each of the delegates got a beautifully bound copy of the Year 2000 China Automotive Industry Development Strategy written by the experts from China and Japan. Thirty-six Chinese and Japanese experts, including He Shigeng, Masuda Masao, and Arakawa, took more than a year to present a report of more than 400,000 words with attachments.

The report classified automobile industry development into four types: the indigenous development in the US; the free trade model in Japan and Western Europe; the dependent or semi-dependent reliance on foreign capital and technology in newly emerging countries such as Brazil; and the indigenous, semi-closed exchanges between the Soviet Union and the members of the Council for Mutual Economic Assistance (CMEA or COMECON)[5]. Experts suggested the Chinese auto industry follow a path of indigenous and open development.

The Chinese and Japanese experts analyzed the shortcomings of China's auto industry at that time.

The Chinese automobile industry had basically taken a closed development path since its inception and was affected in three ways:

— The first was the Soviet model, the fully closed, universal production system of Soviet industry;
— The second was China's old economic management system, with too much centralized state control and too much control over companies;
— The third was the lack of industrial policies, research on rational structures and development of key and strategic industries.

---

[5] The Council for Mutual Economic Assistance (English abbreviation COMECON, CMEA, CEMA, or CAME) was an economic organization from 1949 to 1991 under the leadership of the Soviet Union that comprised the countries of the Eastern Bloc along with a number of communist states elsewhere in the world. COMECON was the Eastern Bloc's reply to the formation in Western Europe of the Marshall Plan.

The automobile industry had long been the same as ordinary processing industries. This had delayed its development.

The Japanese experts made the following assumptions about China's indigenous, open development path, based on experiences of the automobile industry in other countries:

— The greatest advantage would be the huge potential of the domestic market. China's auto industry would not be like in other developing countries. It would follow the path of a big country like the US.
— Technology and capital had to be brought in to develop the industry. Japan's experience could be drawn upon for this. Brazil and South Korea had too much dependence on foreign technology and capital. That would not apply to China.
— Countries which began automobile production at a later stage could develop with government protection and support. This was an effective method adopted by all countries.
— Experiences of the automobile industry in various countries proved that it could not become a strategic industry without competition. Therefore, creating a competitive environment was one of the main features of an indigenous, open development path.

Experts predicted that the proportion of official cars and taxis in car ownership in China would fall from 100 percent then to 47–57 percent in 2000. Thereafter, this demand would be saturated and the dominant domestic market would become private cars.

Thirty years later these assumptions were all true.

Rising stars need policies, not money

The feasibility meeting at Shiyan solved two problems: whether or not China would make cars and how they would do it. Everyone agreed they would be made, but there were different proposals on how it would be done.

Some departments proposed implementing a unified national plan which relied on state investment. The nation should build a new factory, the Third Automobile Works, specializing in sedans. Another factory would be built 3–5 years later.

The factory directors of FAW and SAW were two rising stars among the experts at the conference. Geng Zhaojie of FAW and Chen Qingtai of SAW were both about 50 years old. They would take the automobile industry to a new level.

Geng and Chen held similar views. They did not agree with the CNAIC plan for a stand-alone car factory, but proposed that it should be done within FAW and SAW. Policies were needed from the State, not money.

Output of 300,000 per year was the starting point for economies of scale at foreign car factories. Geng Zhaojie had another view on this. He said that economies of scale of start-ups were different. Where funds were not in abundance, as in China's case, the scale at the start could be smaller.

FAW had brought in a Chrysler 2.2-liter engine production line from the US to develop light trucks. It was preparing to bring in an old Dodge 600 sedan production line for the new generation of the Red Flag. I interviewed Geng at the meeting. He showed me exterior designs for the new Red Flag. He said that most of the Dodge molds could be used, as long as the front and rear were changed. The investment would be smaller that way and it could be produced more quickly. These high-class, official sedans could be manufactured quickly to replace imports.

Chen Qingtai, the scholarly SAW factory director, reminded everyone that the unconditional state support of FAW and SAW no longer existed. Building a new factory would not be difficult, but it was impossible in a short time. Aging of the new plants would be a problem with constant changes of models in the international market. The Soviet Union built the Tolyatti[6] car factory. Sixteen years after it was finished,

---

[6] Tolyatti, also known in English as Togliattigrad and Italian as Togliatti in honor of Palmiro Togliatti, is a city in Samara Oblast, Russia. Internationally, the city is best known as the home of Russia's largest car manufacturer AvtoVAZ (Lada), which was founded in 1966. It was previously known as Stavropol-on-Volga (until 1964).

they had to ask foreigners to help with their first change of model. That should be a lesson for us.

Therefore, he advocated using capabilities of existing enterprises to build two or three car groups.

Chen invited me to his home one morning for a chat. He told me SAW had imported several popular foreign models as prototypes. The factory leaders and product engineers were driving them on Sundays to decide which model was most suitable for China's road conditions.

We spoke about the money needed to establish the industry. Chen argued that it was no longer possible for the government to provide the money. The investment system had to be restructured, and with state support, enterprises should raise funds by accumulating retained profits, forming stock companies, and taking out loans. The domestic financial market was opening up, and the global automobile industry was moving to the third world. The best strategy was for companies to manage debt and undertake modern car production.

Their plans were very clear. FAW would start with a 30,000-vehicle pilot plant to replace imports, beginning with higher class cars and

First Director Rao Bin and new young Director Chen Qingtai rejoiced as FAW passed state inspection in 1985.

developing downwards, then building a light car factory with an annual output of 150,000 vehicles, and finally achieving a capacity of 300,000 vehicles. SAW was inclined to an export-oriented direction, planning to gradually reach 300,000 vehicles.

I fully agreed with the views of these two rising stars and immediately recorded them for Xinhua's internal reference. The details of their conversations are still there today. It is remarkable that their ideas became reality.

Beidaihe: The car industry gets its birth permit[7]

Chen Qingtai, bringing two or three people with him, visited me at Xinhua News Agency one day in early August 1987 as I was about to end work.

"I don't have much time and I have to leave soon, but I've come to ask for your help," he said. He had just returned from the US and had heard that Central Government leaders would be meeting at Beidaihe to discuss the development of cars. He still didn't know if SAW had been designated to produce them, so he was eager to attend. He was going to Beidaihe with his secretary Ma Yue that night and hoped I would give him some extra ammunition by writing an internal Xinhua reference document stating his views on developing small cars for export.

"For better or worse, cars have become a symbol of human civilization in the 20th century. But in China we're still worrying about whether we will be allowed to produce them," said Chen Qingtai. "When I rode in a car across Manhattan and saw the traffic flow by the window, my heart skipped a beat. Why can't a country with a 5,000-year-old civilization do that?"

"I have noticed that although the US has the world's top automobile market, third world products can still enter it. The South Korean

---

[7] Under China's one-child policy (1979–2015), permission was legally required before having a child. The permission was given in the form of a family planning service certificate which was checked when a woman went for a pregnancy test, child delivery, maternity insurance, etc.

Hyundai Pony[8] and the Yugoslavian Yugo have captured a large share of the small car market with low prices and reasonable quality."

"This is inspiring. With the internationalization of the automobile industry, production of small, lower profit cars is shifting to the third world where labor is cheaper. This transfer of funds and technology wasn't possible in the past and it's an opportunity for us. Third world auto products can enter global markets, as long as they meet international standards. As China becomes more open, its cars must gain a foothold in the international market in order to be competitive in the domestic market. Therefore, SAW will work to become an export-oriented manufacturer of medium and small displacement cars."

After speaking for an hour, he stood up to say goodbye. I told him I would put it together and send it to the decision makers before the meeting.

As we shook hands, I hoped he felt my heartfelt wishes for SAW and our dream of a Chinese car.

Chen and Ma drove through the night to Beidaihe in SAW's Beijing office Toyota Crown. It was raining heavily and the car got stuck in the mud a few times, so the two Tsinghua University graduates and automobile professionals had to get out and push.

The car climbed in and out of quagmires.

The three- or four-hour journey took 15 hours.

On the afternoon of 12 August, blue skies followed the rain, and the waves were lapping the shore.

In a tree shaded villa on the Beidaihe coast, Deputy Prime Minister Yao Yilin, who was heading the meeting, together with Li Peng and Zhang Jinfu heard the report from Chen Zutao, Chairman of the China National Automotive Federation (CNAF), on developing cars.

Chen Qingtai made it to the meeting and had in depth conversations with the leaders on SAW's car development, through the good offices of CNAF Vice Chairman Bo Xiyong.

---

[8] The Hyundai Pony, also known as the Hyundai Excel, Hyundai Presto, Mitsubishi Precis and Hyundai X2, is an automobile which was produced by Hyundai Motor Company from 1985 to 2000. It was the first front-wheel drive car produced by the South Korean manufacturer.

State approval of FAW's project seemed to be a foregone conclusion, but SAW had to push its way onto the national plan.

Chen Qingtai reported on SAW's plan. Its target was a model with an engine displacement of 1.3–1.6 liters and the largest market volume. That type of car could be used as the main domestic official vehicle in the future and could also become a leading export product. The scale of production would be 150,000 per year in the first phase and 300,000 in the second phase. He proposed two ways to proceed. The first would be by introducing technology, building their own factories, substituting for imports, and exporting in the long-term. The second would be through joint development, constructing joint venture factories, being export-oriented, and by substituting for imports.

Chen's presentation won the approval of the leadership.

State Councilor Zhang Jinfu said, "SAW's cars must be built with new technology. The technical threshold has to be higher, and the most important thing is to have a new engine. The basic parts should make full use of FAW and SAW, and they must be made in large quantities. The production of cars by SAW can start with parts and components. Foreign exchange funds will be easily obtained by making buy-back a priority."

Li Peng said, "SAW's car has been fully discussed and is ahead of both FAW and Shanghai in terms of model type and engine class. I agree with their ideas in principle. Cars will come from these three entities. Let's not be reckless and build a fourth or a fifth. That would only bring more problems."

Yao Yilin concluded by saying, "SAW's plan is good. I agree that it should aim at exports. For sedans we can settle on Shanghai, FAW and SAW, but no others. FAW's project documents are being processed, and SAW can go ahead, too. Their project has to be set up before they can take the first step."

A major decision was made at Beidaihe that day. After all its difficulties, the car industry finally got its birth permit.

After many State Council appraisals, Chinese leaders had finally reached a consensus on the car industry.

Hu Qili, Secretary of the CPC Central Secretariat[9], heard about the report. He was concerned about political stability when the campaign against bourgeois liberalization and the student demonstrations occurred at the end of 1986 and CPC General Secretary Hu Yaobang was replaced. He said, "Automobile imports have become a political issue involving our national self-esteem and self-confidence. India primarily uses domestically manufactured cars. The development of cars must be strategic. There has to be competition, and support has to be laid out strategically on that basis."

Hu Qili wisely warned not to lower the quality of cars sold domestically, when he heard people advocating building two cars, one for export and one for domestic use. "That would give us a bad reputation and a loss of prestige internationally. They are all customers, whether foreign or domestic, and they all want high quality. China can make good things if we set our minds to it. Comrade Deng Xiaoping made a good point recently: 'Quality represents the character of a nation.'"

The State Council General Office drafted a summary of the Beidaihe meeting. Article Four said, "FAW and SAW would be relied upon for sedan production in the future. Additionally, Shanghai Volkswagen must get localization done first. No more car production plants would be set up."

That was the original framework and the origin of the "Three Majors" in the "Three Majors and Three Minors" plan. Although Shanghai Volkswagen was not one of the mainstays, its production capacity was still 30,000 vehicles per year.

The so-called Three Minors, Beijing Cherokee (Jeep), Tianjin Daihatsu, and Guangzhou Peugeot, were not mentioned in the summary. Subsequently, the State Council issued a notice emphasizing

---

[9] The Central Secretariat of the Communist Party of China is a body serving the Politburo of the Communist Party of China and its Standing Committee. The secretariat is mainly responsible for carrying out routine operations of the Politburo and the coordination of organizations and stakeholders to achieve tasks as set out by the Politburo. It is empowered by the Politburo to make routine day-to-day decisions on issues of concern in accordance to the decisions of the Politburo, but it must consult the Politburo on substantive matters.

strict control over car production. Apart from the six car factories that had been approved, no new plants would be built. This Three Majors and Three Minors arrangement remained in place until the 21st century began.

2. Volkswagen counterattacks Chrysler

For the 2nd generation of the Red Flag

I interviewed Factory Director Geng Zhaojie in Changchun 20 years later. As my old friend recalled the past, he seemed to regain his enthusiasm. He recalled that after decentralization in 1984, FAW proposed producing 200,000 vehicles a year. This included 100,000 medium-sized trucks, 70,000 light trucks, 10,000 heavy trucks, and 20,000 medium- and high-class sedans. But the planning authority didn't give its approval, mainly because sedans were still taboo then.

However, FAW didn't give up and came up with a different scheme — changing the plan to 100,000 trucks and 100,000 light vehicles. Sedans were hidden under the name of light vehicles. The new plan was soon approved.

Geng became the manager of the main FAW factory in June 1985. The first major thing he had to do was produce sedans disguised as light vehicles. A full 10,000 mu[10] of land was soon requisitioned and the preliminary civil works began.

How big was this land? When Rao Bin was the factory director at FAW in the 1950s, the area of the factory was almost the same as that of Changchun's old city, so it was dubbed Rao's Half City. Because of the large land acquisition for the new factory, it was likewise called Geng's Half City.

Geng, who was already a deputy in the National People's Congress, used to stand on the steps of the Great Hall of the People when he went to Beijing for meetings. When he saw the dense mass of imported cars, he became determined to restore the Red Flag sedan.

---

[10] Mu is a Chinese unit of land measurement that roughly equals 797.37 square yards (0.16475 acre, or 666.7 square metres). 10,000 mu equals 1647.45 acres or 666.7 hectares.

Geng came up with a strategy for moving their light vehicles to other places through joint operations. They were put into local companies in Jilin, Harbin, and elsewhere, giving them more room to produce cars.

The first-generation Red Flag sedan was a product of the Great Leap Forward in 1958, when everyone was in a state of confusion. They were melting their pots and pans to make back-yard steel in homemade furnaces and competing for record grain production all around the country. Falsification and exaggeration were causing uproar. When the dust settled, everything lay in ruins. But the Red Flag sedan was one of the few positive outcomes.

Discontinuing the Red Flag sedan was painful for the people at FAW and resuming production of a second-generation Red Flag was something they had been longing for. It was difficult for outsiders to understand that kind of devotion.

When the Red Flag was discontinued in 1984, FAW had developed two new models, the 750 and 760. These two cars represented a transformation. It was downgraded from a high-end to a middle-to-high grade vehicle, becoming a car for official use. Of course, the large, bullet-proof Red Flag would be retained for the nation's leadership.

The 750 and 760 were called the Second-Generation Red Flag. Geng Zhaojie said three prototypes were made. Prior to this, FAW also modified the big Red Flag 770G. Standards were relatively high because people were open-minded and key components were being imported. When the 770G was sent to Beijing, General Secretary Hu Yaobang said, "I raise both hands in favor of restoring production of the Red Flag."

In the face of a tide of imports in 1986, the Central Government also began to think about resuming production of the Red Flag.

Geng Zhaojie knew it had to have a good engine, whether it was a car or a light vehicle. If engine factories were built from scratch, three to five years could be wasted and the opportunity lost.

Chrysler had a production line for 2.2-liter engines in Mexico with an annual output of 300,000. It was brought in by Volkswagen during the 1970s oil crisis and when the crisis passed it was idled. Geng flew to Mexico and was satisfied with what he saw. He bought it immediately because its products could be installed in both light trucks and cars.

FAW Plant Director Geng Zhaojie and VW Chairman Carl Hahn laid the foundation for FAW Volkswagen's success with their strategic vision.

Coincidentally, Chrysler also had a Dodge 600 car production line it was preparing to eliminate. And it so happened that the heart of that car was the 2.2-liter engine. FAW envisaged buying both of these production lines and re-creating the little Red Flag. On his way to Shiyan to attend the automobile seminar, Geng made a special trip to Beijing to report to State Machinery Industry Committee Chairman Zou Jiahua and Vice Chairman He Guangyuan. There were growing calls to resume producing the Red Flag. Zou suggested resumption of the Red flag along with the importation of the two production lines.

Hahn: A Long March in winter

Volkswagen Chairman Carl Hahn, who was on vacation overseas, received a fax in autumn 1987 from his friend Walther Kiep[11] informing him of FAW's project to build 150,000 sedans.

---

[11] Walther Kiep (5 January 1926–9 May 2016) was a German politician of the Christian Democratic Union (CDU). He was a member of the Bundestag between

Hahn later recalled, "That fax profoundly motivated Volkswagen. I immediately wrote a letter to Director Geng Zhaojie and received his invitation to visit within 24 hours."

After the Frankfurt Motor show at the end of September, Hahn and his Chinese affairs group visited Beijing on the way to Changchun on 20 October. Hahn recalled the harsh environment at FAW then. They didn't take off the military coats that their hosts provided them for 24 hours. It was bone-chillingly cold everywhere except in the factory's electronic data room.

Before Hahn arrived, Beijing called to tell Geng that Hahn's visit was only a courtesy call and not to have any substantive negotiations. But that didn't seem to stop Hahn and Geng.

They began intensive discussions as soon as he arrived. Dinner was a steaming hot northeastern dish. After dinner, Hahn had a look around the factory. "I saw the whole assembly line, from the foundry to the machine shop. I saw the work that had been done for the new factory building. My God, why didn't I know sooner that there was such a big car factory in China," Hahn said in surprise.

In 1985, Shanghai Trailer Co. proposed producing the Audi 100, Audi's top product, in Shanghai. After repeated negotiations between the Chinese and the Germans, it was finally decided to produce the Audi with semi knocked-down (SKD) parts. In the next two years, a total of 100 Audi 100s were assembled by Shanghai Volkswagen for Central Government departments and Shanghai officials to use.

In Europe and in Beijing, I have had long talks about the past with Hahn, who is no longer working. Hahn said that the original intention for introducing the Audi 100 to China was his multi-brand strategy. He smiled and said, "When you go fishing, you will always catch more fish with two poles than with one."

However, Hahn felt that the site was too cramped, so he came up with another idea. He was eager to expand results in China, so he

---

1965 and 1976 and again from 1980 to 1982. From 1971 until 1992, he was treasurer of his party at the federal level. In this position, Kiep installed a system of unreported income accounts, leading to the CDU donations scandal in 1999.

proposed moving Audi production from Shanghai to FAW in Changchun. Extensive cooperation with FAW began with assembly of the Audi 100.

Geng was being taken advantage of by Chrysler at that time and Hahn's proposal was exactly what Geng had in mind.

When Chrysler learned that the Chinese government would give approval to FAW to produce cars, it immediately raised the asking price for its decommissioned Dodge 600 assembly line. It proposed a down payment of US$17.6 million, and couldn't guarantee when the tooling dies would be transferred. The negotiations became difficult. FAW had bought Chrysler's engine technology and equipment, and it seemed the engines they produced could only be installed in Dodge models. Chrysler thought it had FAW in its pocket.

VW's suggestion gave Geng an extra trump card now. But he also felt in a bit of a bind because it was too late to change the Chrysler engine. He told Hahn if he worked with Audi, it was important to put Chrysler's engine in the Audi 100. This would normally not be acceptable, and he didn't expect Hahn to answer straight away.

Hahn said to Geng, "I'll have a satisfactory answer for you if you come to Wolfsburg in four weeks' time." Hahn had done his homework beforehand. He knew that Audi technology was used in the Chrysler engine and that Audi was very familiar with it.

"Negotiations carried on until 1 am," Hahn recalled. "I talked about localizing in my opening remarks, and we both expressed interest in setting up a joint venture. I pointed out that we could provide a complete die set for the Audi 100 from our South African factory soon and at a preferential price. We also discussed the possibility of producing the Volkswagen Golf and the Chinese side expressed interest."

Cars were political products in China. Whether you could make cars and with whom you could cooperate were not decisions a factory director could make. Moreover, cooperation with Chrysler seemed to be a foregone conclusion. However, Geng Zhaojie always did things that others didn't even dare to think about or do. He called Zou Jiahua, Chairman of the National Machinery Industry Committee, and asked if he could play two cards at the same time.

Zou, who was also a risk taker, let Geng make decisions as the situation required.

Geng was back in Changchun a month later. Chief Engineer Lin Ganwei and Chief Economist Lü Fuyuan headed up two separate groups: one flew to Wolfsburg and the other to Detroit.

Hahn surprised Lin in Wolfsburg. A modified Audi car with a 2.2-liter Chrysler engine was shown to him. It even had the FAW badge at the front. This showed how efficient the Germans were and how sincere they were.

Lin Ganwei later recalled, "It surprised us a lot. All the connectors were forged and the entire assembly was good. But the 2.2-liter engine was relatively tall, so the front of the engine hood bulged, like the front of a 747 aircraft. Highway tests were carried out at a maximum speed of 205 km per hour. All aspects of its performance were satisfactory."

The delegation liked the Audi. Its body design was a generation ahead of the Dodge 600, and it was roomy and comfortable. It was very popular and had won international awards. It was versatile like the Shanghai Santana and was suitable for component localization.

Meanwhile in Detroit, Chrysler still refused to yield on the price of the old Dodge assembly line. CEO Lee Iaccoca arrogantly refused to meet with Zhu Rongji, Deputy Director of the State Economic Commission and Chen Zutao, Chairman of the China National Automotive Federation when they came to facilitate negotiations.

Chen Zutao offered Chrysler US$1 million for the old production line, only a fraction of the asking price. He said bluntly, "This old line is to be dismantled. We won't buy it. Its value is zero. It's better for you if we buy it than it is for us, but we won't pay a high price for it."

Zhu Rongji hinted to Chrysler that they had sent a delegation to Volkswagen to compare offers. However, Chrysler turned a deaf ear to his message. The Chinese left disdainfully and decided to cooperate with Volkswagen Germany. When Chrysler learned the Chinese had chosen Volkswagen, it made a last-ditch offer, reducing its price to one US dollar!

But it was too late. FAW and Volkswagen signed a letter of intent to produce Audi sedans a few days later.

Experts have their own strategies for success

The government wanted to re-launch the Red Flag and Audi's unexpected participation was a lucky break. Geng Zhaojie said that there are two ways to bring it back into production. One was to restore the original Red Flag, but they didn't have the technology to do that. The other was to combine it with the Audi and the timing was perfect for that.

FAW wanted to raise its annual output to 150,000 cars, but lack of funds was a major limiting factor. The government's policy was to assemble 30,000 Audis first, the sooner the better. The State would collect taxes only, leaving the profits to the company. The number eventually grew to 150,000 vehicles from 30,000.

Geng Zhaojie said, "It was like an aircraft taxiing before it takes off. Assembling 30,000 Audis was a pilot project to obtain technology, capital, and experience."

Dr. Hahn didn't see the 30,000 sedans as a pilot project at all. He had his eye on the 150,000 vehicles.

Geng said, "We were both fishing. I took Hahn's bait, made concessions, and caught a big fish. He also took my bait. Knowing what he wanted, I forced him to give way even more on the 30,000 Audis."

The 19 million Deutsche Marks technology transfer fee was set aside. If an agreement was reached for the 150,000 Golf cars by 1991, the fee would be exempted. 10 million of the 21 million Deutsche Marks for the dies would be paid in advance and the rest would be invested in future joint ventures.

Perhaps the two savvy entrepreneurs both achieved goals and successes. Geng Zhaojie shrewdly bought Audi's complete vehicle technology for only 10 million Deutsche Marks, including nine complex technologies, the personnel training, and the dies. FAW at last had a good partner and was soon able to get up and running. Hahn was able to get Volkswagen a monopoly over two of the major producers, FAW and Shanghai, in China's Three Major and Three Minor car projects — foreshadowing the capture of 50 percent of the domestic automobile market by FAW and Shanghai in the 1990s.

Geng and Hahn became close friends in the process. Geng was invited to Wolfsburg in August 1988 and Hahn met him on arrival. He gave Geng a gift, a matchbox-sized camera. Hahn said, "You can take photos anywhere in Volkswagen. You are our friend."

With goodwill on both sides, the negotiations were completed in four or five months. That was fast compared with the six years it took for Shanghai Volkswagen.

By the time the thousands of tons of drawings and materials arrived from Wolfsburg three months later, the equipment for FAW's 30,000 car pilot plant was already being installed, preparations were underway to produce 150,000 cars, and the first batch of newly assembled cars was coming off the assembly line.

When Hahn went back to Changchun for the ribbon-cutting ceremony, Geng presented him with two tiger-patterned tapestries. "There are a lot of allusions in Chinese. Do these tigers mean anything?" Hahn asked. Geng answered in eight words: "A tiger coming from the mountains is unstoppable."

Twenty years later, Geng Zhaojie talked about the introduction of Audi technology. The intent was to build up their own brands while strengthening their international brands. He said that Openness was a firm, long-term policy, but it needed joint ventures and indigenous brands that could stand on their own. "We didn't think one would get more than the other," Geng exclaimed, "but now it seems that Hahn got more."

3. A new format for China's Three Majors and Three Minors

FAW-Volkswagen gets the upper hand

Coinciding with a severe economic adjustment, the domestic auto market fell into a slump in 1990 when products were heavily overstocked. The Political Bureau of the Central Committee issued a decree restricting the import of small cars, forcing members of the Politburo, the Secretariat, and the State Council Standing Committee to use domestic cars. FAW was in luck. Only the Audi met the requirements.

FAW didn't relax because of strong demand for the Audi. It had the big fish — a joint venture project with Volkswagen to produce 150,000 popular Golf and Jetta sedans.

When it learned that Volkswagen had a discontinued Golf production plant in Westmoreland, Pennsylvania, FAW sent Chief Economist Lü Fuyuan and his assistant Li Guangrong to Wolfsburg to negotiate its acquisition. Volkswagen's asking price was US$39 million, but FAW had only managed to raise US$20 million. After 21 days of discussion, VW dropped its price to US$25 million, but they still couldn't reach an agreement.

At the farewell banquet, Lü Fuyuan, who had a good command of foreign languages, listened as the Germans chatted with each other about losses and layoffs because Audi production had been unable to reach a break-even point. He remembered that when he went to the State Planning Commission before he left China, he heard that 20,000 Audi knock-down kits would be imported for government officials' cars. He used this information as a bargaining chip and asked VW at

The Jetta production line brought from the US set the stage for FAW-Volkswagen.

the banquet table, "Will you give me the Westmoreland equipment if I buy a lot of your Audi kits?"

So the two sides reopened negotiations. The agreement they reached was very different from where they had started. FAW purchased 14,500 Audi kits, and Volkswagen gave assembly line equipment for annual production of 300,000 Golf cars to FAW for free. FAW only used 7 percent of its funds to buy the factory with this discount.

Lü Fuyuan flew to the US immediately and organized Chinese graduate students to help take possession of the factory. The Americans were dumbstruck. More than 100 FAW employees subsequently arrived in Westmoreland to dismantle, number, pack, and transport the 10,000 tons of equipment for reassembly back in Changchun. It was to become the most advanced car assembly plant in China. Later, it became part of the FAW-Volkswagen joint venture and the production base for the Jetta.

Lü Fuyuan served as the first Minister of China's Ministry of Commerce[12] 13 years later.

FAW and Volkswagen signed a contract for the joint production of Golf and Jetta sedans at the Great Hall of the People on 20 November 1990. FAW-Volkswagen was established the following year. The ratio of shares between the parties was 60:40. Audi joined FAW-Volkswagen in November 1995, and the share ratio was changed to 60 percent for FAW, 30 percent for Volkswagen, and 10 percent for Audi.

The first Jetta cars produced by FAW-Volkswagen were unveiled in Beijing on a warm and sunny winter day at the end of 1991. I was invited to do a test drive, along with a female reporter from *Economic Daily*. The car had a test drive license plate and it was stopped by a policeman on Chang'an Avenue. I told him that it was a domestically produced car that FAW had just brought out and I wanted to drive it past Tiananmen Square. The policeman looked at the car inside and

---

[12] In the spring of 2003, the former Ministry of Foreign Trade and Economic Co-operation (MOFTEC) went through a reorganization and was renamed Ministry of Commerce.

out with great interest and finally gestured for us to pass. Back at Xinhua, I sent out a report, "The First Domestically Launched Jetta Sedan Drives Past Tiananmen Square," announcing the launch to China and the world.

Setbacks for Shenlong (Dongfeng Peugeot Citroën)

Chen Qingtai spoke with policy makers at Beidaihe and SAW obtained its status as one of the Three Majors of the car industry. SAW dispatched personnel to inspect a number of car companies in Europe, Japan, and the US. Citroën in France was the most cooperative. The company's president said that they would close their international department if discussions with SAW didn't succeed. Citroën agreed to cooperate with SAW to produce the new ZX, which was to be launched in 1991. The French government offered loans and gave 20 million French francs to fund a feasibility study. SAW and Citroën signed an agreement for a 300,000-car joint venture in July 1988.

The Chinese government approved their cooperation at the end of 1988.

Negotiations were in full swing in the first half of 1989. Ma Yue, Song Yanguang, and Song Zuwei flew to Paris in late May and negotiations went on smoothly. The ZX was expected to be launched in both China and France in 1991.

However, an unexpected political incident suddenly occurred on 4 June, just two days before the signing of the contract. France joined the West in imposing economic sanctions and announced that it would stop high-level contacts between the two governments. Despite the thorough preparations, the SAW Citroën project was put on hold.

It was sidelined for three years. After numerous disappointments and struggles, SAW, renamed as the Dongfeng Group, signed a contract with Citroën to establish the Shenlong joint venture on 19 May 1992. Dongfeng held 70 percent of the shares, Citroen held 25 percent, and between them two French banks held 5 percent. It became the largest joint venture project in China, with a total investment of 10.3 billion yuan.

The first Fukang[13] cars didn't come off the assembly line in Xiangfan, Hubei Province, until 8 September 1995. The Shenlong Company moved from the Shennongjia area near Shiyan and settled in Wuhan, 500 kilometers from the parent company. The biggest and most advanced car factory of the day was built in the Wuhan Economic and Technological Development Zone.

The Fukang was the first Chinese branded joint venture car. The 1.36-liter Citroen ZX sedan that came out in the 1990s was the chosen model and it was in step with the French technology. It had a small engine, low fuel consumption, and was fast and smooth. It combined a set of new 1990s technologies, such as its teardrop shape for lower wind resistance, passive rear wheel steering[14], and anti-collision body structure.

Hatchback models were not widely accepted in the restricted family car market in China. Dongfeng-Citroën had difficulties from the start. However, the Fukang was the only car that could be driven when Wuhan suffered a flood in 1998. It was the first to receive a green environmental label in Beijing in 1999. And it was the first in China to pass a safety crash test. These attributes gave the Fukang a good reputation.

Project No. 1 for Shanghai Volkswagen

Shanghai Volkswagen was doing well in the early 1990s. One hundred thousand Santana sedans had been produced by 1992. Localization was more than 80 percent, but supply didn't meet demand. As it began to make money, Shanghai Volkswagen hired the first foreign coach for the Chinese national football team, Klaus "Uncle Shi" Schlappner[15].

---

[13] Re-badged Citroën ZX model.

[14] Passive rear wheel steering uses the lateral forces generated in a turn (through suspension geometry) and the bushings to correct this tendency and steer the wheels slightly to the inside of the corner. This improves the stability of the car through the turn.

[15] Klaus Schlappner (born 22 May 1940) is a football manager. He is predominantly remembered for his first spell with SV Waldhof Mannheim, where he led them to

The car industry makes a lot of money, but also spends a lot. SAIC made a bold decision to discontinue its Fenghuang brand sedan which had been in production for 34 years and use the funds to increase production at Shanghai Volkswagen. Its factory was merged with Shanghai Volkswagen. Regrettably, they had no other choice. There were several versions of this story.

The Sino-German joint venture provided more capital, investing 2.5 billion yuan to begin the second phase of Shanghai Volkswagen's transformation. The Shanghai Automobile Factory was converted to a second Volkswagen plant with an annual output of 90,000 vehicles. It was later identified as Project No. 1 of which the Mayor of Shanghai was personally in charge. Under China's new economic structure, local enterprises fully backed by a city had shown obvious development advantages over state-owned enterprises.

This second factory was built to put a new model into production. The Santana's rear seat space couldn't meet the needs of officials, so a new model was put on the agenda.

Shanghai Volkswagen sent a group of nine people, including Qin Zhongnian, to Sao Paulo, Brazil in March 1992 to be part of Volkswagen's and Autolatina Brazil's joint development team. The goal was to design a new Santana with a longer body and more comfort to satisfy the demands of the Chinese market. It was based on Autolatina's Spruce sedan, which was also on a Santana chassis. Even though there was more observation and learning than hands-on design, it was still the first time China's auto industry had participated in the design and development of an international car.

The Santana 2000 debuted at Shanghai Volkswagen's 10th anniversary celebration at the Longbai Hotel on 10 October 1994.

The Santana 2000 was the first car I test drove in China. I followed the Silk Road from Lanzhou to Dunhuang across the vast expanse of the Gobi Desert. Chasing the floating mirages on the horizon of the

---

the 2. Bundesliga title and several seasons in the Bundesliga as well as being the first foreign coach to manage the Chinese national football team.

Great Gobi in the roomy, comfortable, and powerful Santana 2000 was one of the most memorable driving experiences of my life.

Shanghai Volkswagen had already left the original 30,000 limit far behind, going on to reach a capacity of 200,000 vehicles. It fully deserved to be one of the Three Majors.

The comfortable lives of the Three Minors

The Three Minors specifically referred to three local projects — Tianjin Xiali, Guangzhou Peugeot, and Beijing Jeep — in the 1990s.

Tianjin introduced Japan's Daihatsu technology to produce mini-vans in 1984. Two years later, the 1-liter Xiali mini-car that Daihatsu had just put into production was introduced. Deng Xiaoping praised it and Xiali progressed to become one of the Three Minors in 1987. It had a production capacity of 50,000 vehicles by 1994.

It was the only car that was not the product of a joint venture.

The Xiali was a popular three-cylinder car without technical sophistication. It broke down a lot and wasn't as comfortable or stylish as the mid- to high-end cars. However, Xiali was a step closer to being part of the public car culture because it cost less than 100,000 yuan. It was popular because it was fuel-efficient, small, and maneuverable, and its maintenance cost was low. It took the top spot among Chinese taxis, with a 39 percent share.

The Guangzhou Peugeot 505, a seven-seater station wagon was popular for a while in the early 1990s. I interviewed its French General Manager, Meng Gaofei, who liked paragliding and fast cars, and its Chinese Deputy General Manager, the refined and cultivated Liu Yuwei. They both became friends. Meng complained that their main product, the Peugeot 505, was larger than a Santana and had a more advanced configuration, but the price set by the government was lower than the Santana.

I was impressed by the bright wavy patches on the walls at Guangzhou Peugeot. With its bright architectural exterior, it wasn't dull and oppressive like most factories. It was the first in the domestic auto industry to do multi-vehicle mixed production of light trucks, sedans, and seven-seater station wagons and the first to do export buy-backs[16].

However, the high cost of producing multiple models, insufficient investment by the shareholding parties, and a weak domestic parts support system for a 30,000 vehicle production rate, laid the groundwork for its eventual decline.

The Beijing Jeep Cherokee ranked as a high-end model. The SUV concept had not yet taken shape, but there was no shortage of loyal fans in China for the representative of the US Jeep with its off-road performance, comfortable ride, and masculine look.

However, the market, in which official vehicles occupied an absolute position, was affected by fluctuations in national economic policies. The roller coaster changes in the market hurt the company badly.

At the height of the government's administrative remediation in the chilly spring of 1990, the purchasing power of the controlling group rapidly dissipated. Beijing Jeep went from doing booming business to suddenly having few customers. Production capacity had reached 15,000 after five arduous years. However, by mid-April, only 48 localities had gotten approval from the Collective Consumption Control Office to purchase Cherokees. Brand new Cherokees came off the production line every day in accordance with the plan, while at night the car lot at the factory was as quiet as a deserted cemetery. Only a northwest wind swirled among the black shadows of the cars.

A year later the situation had changed 180 degrees, although the government had only slightly relaxed its remediation efforts. The Beijing Cherokee reached its annual production figure just after the first quarter. But the company's other local off-road vehicle, the Beijing 212 which had not been changed for nearly 30 years, was more popular than ever. Beijing Jeep sold 80,000 vehicles in 1995, a peak it never reached again.

---

[16] Counter-trade arrangement in which an exporter of equipment agrees to buy a specified portion of the manufactured goods as an incentive to the equipment purchaser.

02 Iacocca's advice: An impression of Lee Iacocca

At the beginning of Reform, what Chinese people heard about Chrysler probably came from *Turning Defeat into Victory*, the Chinese title for the autobiography of its President, Lee Iacocca. It tells about an entrepreneur struggling against adversity and making a comeback. It was a textbook for the Chinese business community, especially for the newly emerging entrepreneur class. A total of 6.4 million copies were printed in various translations, and Iacocca became an instant celebrity in China.

As President of Chrysler, Iacocca visited China in 1988, causing a sensation in media circles. However, few reporters knew that this astute man had just lost at the hands of the Germans and had failed to gain an important inside track to the Chinese auto industry.

Dozens of newspapers and magazines wanted to interview him. At a press conference, reporters kept asking about Chrysler's cooperation with the Chinese automotive industry. But he avoided a direct answer, although he was happy to talk about funding for diabetes research. One of his close relatives had died of the disease.

Iacocca said, "Today, the 15th of October, is my 64th birthday. I didn't think I would spend the day interacting with colleagues from the Chinese business community at the Great Hall of the People in Beijing. It shows that the world is becoming smaller, and communications are bringing people closer."

"Shall I say something about my impressions of Beijing? Well, Beijing roast duck is delicious, and I've eaten quite a lot of it."

"What impresses me most deeply is the openness in China. When I went to climb the Great Wall, a lot of people recognized me. Students, and even military people, came up to chat with me and talk about my book, and asked me about a lot of things apart from the book. They wanted to know about everything."

"And the leadership has been ready to talk frankly about problems. There are many aspects of our cooperation for which solutions still haven't been found, but at least I've made some friends in the past week."

If I hadn't looked back through my old notes, I wouldn't have believed what Iacocca said 20 years ago, which was:

"The world is getting smaller, but movement towards integration, osmosis, cooperation, and division of labor in the world economy is not happening as people would have wished. Humans have never been so interdependent in terms of raw materials, technology, inventions, and markets. However, competition is intensifying. The good old days of a unified world are gone. There is no other choice now but to bring out the best products. In the past, I thought I would only be competing with GM and Ford Motor Company. I didn't expect to be competing with the Japanese and Koreans. I was wrong. I used to think that the best car designs were always from Detroit. I was wrong. I used to think that countries several generations behind the US could never catch up. I was wrong..."

"It's the Chinese people who'll decide the speed of China's transformation. As a foreign investor, one must put to use that most important lesson of Eastern wisdom — patience. That's difficult for us Americans. Patience isn't something we're good at. When we see an opportunity, we're impatient to act. Chrysler is looking forward to cooperating with the Chinese automotive industry, and I hope we'll be successful. Perhaps I'm saying this today because it happens to be my 64th birthday. Is it because I'm getting old that I want this to happen?"

Iacocca showed some emotion, which puzzled many reporters. But they didn't know the story behind it.

The day after Iacocca's arrival in Beijing, I met Chen Zutao, Chairman of the China National Automobile Federation. I asked him about the ceremony at Diaoyutai that morning, where Iacocca had donated a Dodge sedan for the 1990 Asian Games. I was the only reporter present and I had a strong sense that the ceremony was brief and had been downplayed.

"He hasn't been able to do business in China," Chen Zutao said. "He's just trying to see if he can peddle some of their light aircraft. As far as cars are concerned, we won't be dealing with Chrysler anymore. He's well aware of that, too."

I don't know whether it was because of caution or arrogance. Although Chrysler sold FAW an engine production line, Volkswagen managed to pluck the peach which should have fallen into Iacocca's mouth while the second-generation Red Flag sedan was being planned.

I thought, if they had just gone with the flow and entered the Chinese mid-to-high-end sedan market with their Dodge 600, Iacocca would certainly have added a few more legendary pen-strokes to the history of his struggle — Chrysler gaining entry into the world's most populous market with the lowest rate of car ownership. Because he took his eye off the intense international competition, the Iacocca who had turned defeat into victory yesterday, let the opportunity slip and departed from the scene.

Iacocca came to China just at that uncomfortable moment. Luckily, Chrysler had recently merged with the US company which produced the Jeep, thus becoming a shareholder of the Beijing Jeep joint venture, which gave some meaning to his trip.

At the press conference in the Great Hall of the People, a reporter asked him for his suggestions for the Chinese auto industry with the same frank attitude of his second autobiographical work *Talking Straight*. A meaningful, humorous light flashed behind his glasses:

"If I had chosen journalism as a career instead of industry, most of my books about the 21st century would have been written in China. All of you sitting here are writing the first chapter of that book. China is cautiously entering the world market, and fortunately, it has many good lessons from the past. Its goal is to prosper, without losing control. I have four suggestions for my Chinese counterparts."

"Firstly, don't let other people take away your markets. It's better to invest in technology, repair your roads, and build cars, rather than buying a lot of cars. Transferring technology to China is easier than it was 30 years ago, so China doesn't have to invest a lot and follow the same detours the West went through."

"Secondly, make your own development plan. The US automobile industry used to neglect fuel economy and ignore pollution. It had to make a detour and spend a lot of money fixing these problems. China should find shortcuts in lessons from other countries to avoid repeating the same mistakes."

"Thirdly, concentrate on specialized production. I've seen the 1930s US model in China, where the output was small, and almost all the parts were produced in the same factory. In the US today, 70 percent of the parts are purchased from specialized factories, which have four times as many employees as the automobile factory itself. The auto factory concentrates human and financial resources and does the key part of the work, gaining competitiveness through large quantities and low costs."

"Fourthly, encourage innovation. The world is filled with its spirit now. Chrysler has taken measures to cut paperwork and meetings and encourage innovation, filling the company with vitality and talent."

I met Iacocca three times. He was tall and sturdy, and seemed older than in the photos in his book. His eyes and mouth gave him a uniquely fascinating expression. He was somewhat reserved in his contacts with Chinese officials, allowing himself to be managed by zealous protocol officers, but he was relaxed and unconstrained with Chinese journalists. He was thoughtful, witty, and even satirical.

Iacocca's visit was brief. Before he left, he said he was coming to Beijing to see the Asian Games in two years' time, but he didn't seem to have done that. His lost opportunity in the FAW car project was known by everyone in the Chinese automotive industry. But his recommendations are still true for the industry today.

Chapter 3

The Family Car Makes Its Appearance

The National Automotive Industry Conference met in January 1986. In his concluding remarks, Deputy Prime Minister Li Peng surprised everybody by departing from his script to talk about his expectations for the family car. He said, "People will be wealthy in a few years, and there may be a demand for cars among families who become wealthy first. We should be developing light family sedans." He mentioned a price of five or six thousand yuan. The small Fiat 126P imported from Eastern Europe cost that much.

It was the first time a State Council leader had mentioned family cars. Acting on a hunch, I wrote an article entitled "Small Family Cars Are on the Agenda." Of course Xinhua couldn't publish it. I gave it to *Capital Economic Information Daily*, a subsidiary of *Guangming Daily*, which to my surprise put it out as a headline story. The Central People's Broadcasting Station also broadcast it in their morning prime time news digest. I was criticized as soon as I arrived at the office for disclosing something so sensitive.

Celebrities, film stars, and people who had the social means, could purchase, with approval, used cars that were disposed of by foreign diplomatic missions through the Friendship Stores. At twenty to thirty thousand yuan each, they were the first private cars after the Cultural Revolution. They weren't affordable for those whose wages were only a few dozen yuan per month. And for those who didn't have the right to own them anyway, they could only be a distant dream.

1. Prequel to a dream

Pilot products and problems official cars

Private cars were tightly restricted as symbols of capitalism following the birth of New China in 1949. The foreign capitalist bosses' cars were driven, but less and less, until they completely disappeared during the Cultural Revolution. Renowned Beijing Opera performer Ma Lianliang was humiliated and beaten to death in Beijing in summer 1966. One of his "crimes" was owning a private car.

The counterparts of private cars were official cars, which were allocated according to rank. Provincial ministry level and above rode in Red Flags, and bureau-level cadres in Shanghais. Before 1984, county and team level could only ride in Beijing 212 Jeeps. Private meant capitalist and reactionary and people avoided it in terror. *China Comment*, a current affairs publication run by Xinhua, was still discussing whether farmers could own their own walking tractors in the early 1980s. It was not until 1984 that State Council documents allowed farmers to own motor vehicles as means of production. It was an epoch-making breakthrough.

Chen Zutao, Chairman of the China National Automotive Federation, was the first to advocate family car ownership.

Five or six years into economic reform, cars were still luxury items, not a means of production. Private cars were under especially strict control. Apart from secondhand cars which celebrities purchased from foreign organizations in Beijing and Shanghai, new cars were unthinkable for ordinary people. There was nowhere to buy them even if they had the money. The Ministry of Foreign Trade tested the water by importing a number of minicars from Eastern Europe by barter trade through gray market channels.

For a long time, even talking openly about private cars was off limits.

The State Council did away with the China National Automotive Industry Corporation, replacing it with the China National Automotive Federation in June 1987. Chen Zutao was Chairman of the Board. A historical achievement of the Federation was to promote the establishment of the car industry and then put family ownership onto the agenda.

Chen Zutao was the son of Chen Changhao, one of the early leaders of the Chinese Communist Party and a political commissar in the Red Fourth Front Army. At the age of 11, Chen Zutao was sent from Yan'an to the International Children's Institute in the USSR. He graduated from the Machinery Department of the Bauman Institute of Technology in Moscow, returning to China in 1951. He was soon assigned as a representative in the USSR liaising in the construction of FAW. He started doing technical leadership work in FAW in 1955. He established SAW and became its first Chief Engineer in 1965.

When the China National Automotive Federation was starting up in 1988, economic reforms had entered a crucial stage. A dual-track system of planned and market prices had brought about widespread official profiteering. Raw materials like steel and aluminum were kept in warehouses, while ownership changed and prices snowballed. The State Council relaxed controls on some products and allowed prices to rise, causing unprecedented inflation. People panicked and began hoarding wool, soap, matches, and even oil, salt, soy sauce, and vinegar, causing terrible suffering. Conflicting systems, interests, and development became tangled, bringing about turmoil.

To get the tiger of household savings back into its cage, it was imperative to call for pilot products which were compatible with the country's overall economic and technological standards and could produce multiple benefits.

In the early 1960s, local governments used to inflate the price of ordinary goods and keep the difference for themselves. In the 1970s, watches, bicycles, and sewing machines in the price range of a hundred yuan became the three major items where supply couldn't meet demand. In the 1980s, refrigerators, color TVs, and tape recorders for a thousand yuan were all the rage. In the 1990s, there was an urgent need to find an industrial product at the ten thousand yuan level with high technology content and wide industrial coverage to be a pilot product for the national economy. Attention fell on family cars which were mainly consumed by ordinary people.

At the beginning of 1988, Chen Zutao told me that since Reform, five or six million people had become wealthy. They were together with ordinary people in a narrow consumption band which could only make inflation worse. If 10–15 percent of them bought a car, demand would increase by 400,000–600,000 vehicles, releasing excessively concentrated consumption. Two things had to be done to develop this new market. Firstly, the government had to encourage private buyers to buy cars. Secondly, production of private cars had to be planned in advance. I wrote an article in Xinhua News Agency's *Economic Reference* entitled "Encourage Rich Families to Buy Cars," describing Chen Zutao's thinking.

Family car ownership was the key to resolving another economic phenomenon.

Cars which officials rode in and were bought with public funds made up more than 99 percent of car ownership. Chinese people had taken this as a matter of course for decades. However, it was dragging the economy into a vicious cycle. The State took money from one pocket to make cars, and from the other pocket to buy almost all of them. The more cars they built, the more money they spent buying them. By the mid-1980s, with the rise in the number of official cars and an increase in official grades, it was unsustainable.

Explosive news from the Ministry of Finance was published on the front page of the *People's Daily* at the end of 1987. From January to October that year, institutional consumption across the country increased 20.2 percent over the same period the previous year, giving warning of nationwide over-consumption. It had to be controlled. The purchase of high-end cars was particularly pronounced. Compared with 1981, the number purchased nationally increased 6.2 times, and the amount spent buying them increased 14.5 times. A major problem was that government departments were going after imported, luxury, and super-luxury cars, and satisfying that desire with gigantic amounts of foreign exchange.

Expenditure of public funds on official cars made up 70 percent of the total. The Central Committee responded by prohibiting party and government agencies from exceeding set limits. Cars were listed as the first among 19 goods that were strictly controlled. However, the result was that sales of vehicles by the joint ventures the country had just invested in fell into a hole.

The only way to get out of the vicious cycle was by allowing private ownership.

Joining forces in the mountains

The China National Automotive Federation was looking for a breakthrough for family cars.

Guizhou Province was called Yelang[1] in ancient times. Its isolation cut it off geographically and intellectually and the saying "Yelang boastfulness" was a subject of ridicule. At the start of winter, a long convoy of vehicles drove along the twisting highway. Chairman of the China National Automotive Federation Chen Zutao, Vice Chairman Wu Shizhong, Li Yinhuan, Bo Xiyong, and General Manager of China National Automobile Import and Export Corporation Zhu Boshan had come to Guizhou for a tour of inspection at the invitation of the

---

[1] Yelang was an ancient alliance of tribes first reported in the 3rd Century B.C. in the area of Guizhou Province, active for about 200 years.

Minister of Aeronautics and Astronautics Lin Zongtang. They toured hundreds of aircraft and missile factories in 13 cities and counties, and held talks which went all night.

Their tour wasn't reported in the Guizhou news, although Provincial Party Secretary Hu Jintao met with both Lin Zongtang and Chen Zutao. Few people took part in the talks, which were limited to assistants of the two industry leaders. I was invited to listen as an old friend of the automobile industry.

The world was shrouded in the dark clouds of the Cold War in the 1960s. Relations became hostile between China and the USSR, and the situation in the Taiwan Strait was tense. "Prepare to fight" became the national policy. Military industries were of key importance, strategic industries moved inland from the coast, and the Third Front construction program was in full swing. Mao Zedong said he would lose one day of sleep for every day that the Third Front project was unfinished. All the nation's strength was poured into it, and first-rate talent came to the mountains to live in thatched huts, as one military industrial enterprise after another arose in the Third Front.

Warmer winds blew at the end of the 1980s. Orders for weapons fell, and the Third Front military enterprises couldn't get ahead in the merchandise economy. At many Military to Civil Conversion[2] trade fairs, these cutting-edge technology companies had to support themselves by bringing out noodle-making machines, knife sharpeners, and electric combs.

In the mountains of Guizhou, when we saw the missiles, the flight test runway, the wind tunnel, the measuring instruments, the machine tools, and other first-class equipment hidden deep in the mountains, I knew we had to start turning swords into plowshares.

When the tour was over, Lin Zongtang said the aerospace industries must support their military products with civilian ones. The car was

---

[2] Defense Industry Conversion or Military to Civil Conversion is a process of turning defense production into commercial production. It became a part of the overall economic reforms of the 1980s as weapons production became excess to requirements and many factories faced being shut down if they didn't find new, non-military products to produce.

the best choice, because it could take advantage of the industry's multitasking and integration. He said they would put 50 percent of the industry's civilian production towards cars.

The automobile industry encountered a new force. The aerospace industry could show its excellence with that product.

Chen Zutao advised the aerospace industry not to copy the existing models of the joint ventures, but to produce minicars, which was a new field.

The aerospace industry, the automobile industry, and Guizhou Province formed a joint automobile company to produce auto parts at first, and when conditions were right, to produce minicars. Buildings of the Guizhou Aviation Industry Corporation parts factory were used for the assembly plant.

The Subaru 360 turns swords into plowshares in Tokyo

I flew from Beijing to Tokyo on 7 January 1989. Japan's Emperor Hirohito had passed away that day. There were no flashing neon lights in the streets. All I saw from above was the two streams of white headlights and red taillights.

It was probably the first family car inspection tour sent from China and its object was minicars. The head of the delegation was Zhu Boshan, and the main members were the heads of several important aviation and space bases in Guizhou such as Bases 011, 061, and 010. My status in the group was as a researcher for the China National Automotive Federation. The Japanese companies we visited were Fuji Heavy Industries, Isuzu, and Toyota.

I was impressed by Fuji Heavy Industry Headquarters in Gunma Prefecture, where I met Momose Shinroku, known as the "Father of the Japanese car."

The Nakajima Gunma Aero Plant, where Fuji Heavy Industries is located, was once an aircraft manufacturing plant. Large numbers of the Japanese Naval Air Force's Zero fighter planes used in the attack on Pearl Harbor, and in the war in the Pacific, were manufactured there. Today there is an expanse of tranquil green in front of the building, giving a sense of history.

At 78, Momose was slender and erect, with bright expressive eyes under his silver eyebrows. His dark blue striped suit and gray tie showed him to be a neat, courteous, elderly Japanese gentleman. He insisted on standing to speak, saying, "You gentlemen have come to Japan to look at minicars, but I've heard that there are aircraft design experts among you. That's the work I used to do more than 40 years ago, so we have a lot in common."

Sun Ruisheng, the General Manager of Guizhou Aviation, was an excellent designer. The F7T fighter trainer he designed was a modern jet with real combat firepower which won the highest award in the country.

What concerned Sun was that his annual production capacity of 300 aircraft was down to only a few dozen airframes because of the decrease in military orders.

A cutting-edge technical team with the highest proportion of scientific and technological personnel in the country was concentrated in the mountains between Anshun and Zunyi. Several CEOs in the delegation had the task of launching a new mainstay product.

Talent was also the key to Japan's rapid economic recovery after World War II.

At the end of the war, Nakajima Aircraft Company was broken up into a dozen or so small private companies and converted to peacetime industries. Fuji Motor Co., Ltd., the predecessor of Fuji Heavy Industries, was one of them. The brains of a group of the company's aircraft designers became its most valuable asset.

There was a shortage of public transport in Japan after the war. In 1952, Momose, the former aircraft designer, was ordered to design a 360-cc mini-sedan engine for the Subaru 360.

"A car was still a distant dream for most Japanese," Momose said thoughtfully. "Those were the days when we were eating sweet potatoes and bamboo shoots after the war. Personal transportation was mainly bicycling and you were envied if you had a motorcycle. Our company had just one old-style car, and it burned charcoal. It stopped and started and couldn't go very far. We embraced the idea of a small private car to replace the bicycle. It was a dream most people didn't think would come true in their lifetime — for the whole family to be able go out-doors in any weather. Probably not even one in ten thousand Japanese had that hope, but it was the basis for the existence of our small cars."

Momose proposed the design principles for the new car. Firstly, it could seat four adults comfortably. Secondly, its price would be kept below 400,000 yen, because experts estimated that the baseline for motorization in Japan was 80,000 yen per passenger. The monthly income of college graduates was 8,000 yen. Thirdly, it would be fuel-efficient, with engine displacement of 360 cc, equal to two motorcycles. And, fourthly, it had to be at least as durable and climb as well as a truck or a bus. People would stop thinking small cars were like toys.

"Roads were very bad in Japan after the war. When the US sent an inspection team to Japan, they were shocked by what they saw. Some roads which were called national roads were unimproved and cars couldn't even go 60 kilometers per hour on them. That's what our Subaru aimed at and it was accepted by the general public," he said.

We saw a Subaru 360 in the Fuji Heavy Industries showroom. It was a beige, streamlined sedan. It had a lot of likable features.

Aircraft design played a part in the car's concept. The brake pedal was moved to provide more passenger space. Its unique suspension and rear engine lengthened the interior by 30 cm. Strength was not reduced to save on weight. The body was made of sheet steel. The seat frames were aluminum alloy. Plastic, which was just becoming widespread, was used for the roof panel, and the rear window was plexi glass.

Chief designer Momose issued weight limits to each design team. These were adapted from the aircraft design process. Ten other Japanese companies were designing subcompact cars during that time, but most of them stopped because they didn't get the results they expected. From the perspective of automobile history, many technical breakthroughs were made with the Subaru 360, and it provided lessons for automobile design in the US and Europe.

A great debate about whether Japan should produce cars took place on the radio at the beginning of the 1950s. With the support of the Ministry of International Trade and Industry (MITI), the Officials Economics School, consisting of a group of young people, came out with the slogan "Cars will save Japan." They argued that the government should foster pillar industries to drive economic development, combining the advantages of Japan's labor force with the scientific and technological achievements of advanced countries to create a cycle of imports and exports. That was the only way the economy could take off in a resource-poor island country like Japan.

MITI proposed the concept of a national car with the slogan "A car for everyone" in 1960.

450,000 Subaru 360s were produced between 1958 and 1970. In the 1980s, when the car replacement rate was extremely high, the world's classic car enthusiasts still managed to keep 9,000 of them in mint condition. When Fuji Heavy Industries celebrated the 30th anniversary of the Subaru 360 in Gunma, hundreds of this car came from all over the world, making a grand sight.

After negotiations In Tokyo and Gunma, the Chinese reached an agreement with Fuji Heavy Industries to bring in technology and some equipment to produce auto parts and subcompacts in Guizhou.

The attractive Subaru 360. 450,000 of these were produced from 1958 to 1970.

At the beginning of the 21st century, East German Trabis were still seen on Berlin streets. Citizens had to register and wait years to buy one.

In Italy, the Fiat 500 was like a family member after World War II. Two young Italians drove this one from Turin to Beijing in 2005.

Porche's 1937 design for Volkswagen's Beetle became a national classic. Its shape is still fashionable today.

Citroen's 2CV family car, developed before World War II, survived the war hidden in a hay shed. In the background is an unrestored 2CV.

BMW's Isetta, a mini car produced from 1955 to 1964, was so small passengers had to enter through a door at the front.

The little Satellite that cleaved the sky and was gone

Not many people know about the Satellite subcompact, a domestically produced car — similar in size to the Subaru 360 — which appeared on the streets of Beijing at the end of the 1950s.

The USSR launched the world's first man-made satellite on 4 October 1957, stealing the limelight from the US in the space race, so the word "satellite" became a source of pride for the greater Socialist family. Countless factories, agricultural communes, products, and even newborn babies in China were named "Satellite." It was a good name for a car because "satellite" is a homophone of "mini" or subcompact in Chinese.

The Satellite was the only car developed for ordinary Chinese people during the car craze of the Great Leap Forward. However, it had to circumvent restrictions on households having them.

"Replace man-powered trishaws with minicars," was the directive issued by Premier Zhou Enlai. He inspected and rode in the little Satellite, accompanied by North Korean Premier Kim Il-sung. Zhou stipulated the price of the Satellite, to ensure that as taxis they maintained the trishaw rate of twenty cents per kilometer.

It was developed by teachers and students in the Department of Power Machinery at Tsinghua University. In an era of hyperbole, the newspapers were filled with "new achievements" whose absurdities can still be seen until today. However, the development of the Satellite car was scientific and rigorous. Directed by their teachers, classes were divided into specialized groups for the engine, chassis, body, accessories, and tooling. Seven prototypes were built.

Zhou Enlai examined a prototype in Zhongnanhai and authorized mass production of the minicars in October 1958.

The Erliigou Automobile Repair Factory which was under the Beijing Transportation Bureau and not far from Tsinghua University was the manufacturing base. It was later called Beijing Second Automobile which became famous for producing the Beijing 130 light truck.

Funding was plentiful. Beijing Second Auto had a lot of technical experts who contributed to the trial production. Their technical repair skills made up for the weaknesses in manufacturing, especially early on when there wasn't enough design experience and their machining precision wasn't up to requirements. They got all of the cars started and out of the factory.

The Satellite was the fifth model designed by teachers and students at Tsinghua University. It was 2.7 meters long, 1.25 meters wide, and had a 1.4-meter wheelbase. It had two doors and weighed 440 kilograms. Its top speed was 60 kilometers per hour. The opposing two-cylinder engine had a 410-cc displacement and it consumed 4.5 liters of gasoline per 100 km.

Ordinary people admired the luxury Dongfeng and Red Flag official vehicles, but the Satellite was a car they could enjoy themselves. The Beijing municipal government arranged for Satellites to replace trishaws to transport passengers to and from the newly-built Beijing

Railway Station and Children's Hospital, and persisted in charging twenty cents per kilometer because incomes were still low.

However, it was that unshakable twenty cents that ruined the Satellite. To reduce costs, it had a centrifugal automatic clutch with a belt drive instead of gears as in most cars. Soon after it was put into operation as a taxi, serious quality problems were revealed. The engine frequently malfunctioned and it was noisy. Problems with the belts were even more serious. They kept breaking and the cars kept stopping, worrying passengers more than the trishaws had.

The sixth and seventh models were later developed. The seventh model was based on the East German plastic-bodied Trabi[3] (Trabant also meaning "satellite"), modified with four doors and with a more powerful engine, but it never became a reality.

The Subaru 360, which was also launched in 1958, was the car of its time in Japan, while the Satellite cleaved the sky and was gone.

By the time the Trabi was being copied, funds had been exhausted. This, coupled with the great famine after the Great Leap Forward, left the Satellite only one way to go. Production stopped, but some organizations went on using them for a couple of years.

When I was a teenager, I saw a small mountain of wrecked Satellites piled up in the weeds beside the railway tracks, languishing in the sun and rain behind the Beijing Planetarium. Not a single specimen of the 200 Satellites produced survives today.

The Trabi by the way can still be seen in Germany today and has become a classic. Friends who grew up in East Germany are full of affection for it. It serves as a symbol of a time and a system. Before the merger of the two Germanys in 1989, the greatest desire of

---

[3] The German word trabant, derived from the Middle High German drabant ("Hussite foot soldier"), means "satellite" or "companion." The car's name was inspired by the Soviet Sputnik satellite. The cars were often referred to as "Trabbi" or "Trabi". Produced without major changes for nearly 30 years, the Trabant became the most common automobile in East Germany. It came to symbolize the country during the fall of the Berlin Wall in 1989, as images of East Germans crossing the border into West Germany were broadcast around the globe.

ordinary families in East Germany was to get permission to buy one. They had to apply and wait, perhaps for several years. It was much like the way Chinese people drew lots at their workplaces to be allocated a ticket for a bicycle in the same period. Although East Germany was the wealthiest country in the Eastern European Socialist family, the Trabi was compact and cute but extremely crude, with doors made of cardboard. A German friend said that they didn't even have a radio, and people couldn't talk in the car because of the engine noise.

Thirty years later in the mid-1980s, with the change from military to civil production, Jiangnan Machine Factory, Jiangbei Machine Factory, Qinchuan Machine Factory, Yimin Machine Factory, and other weapons industry companies were mass producing motorcycles, and there was an opportunity to develop a new generation of Chinese minicars.

In the summer of 1986, I reported on an exhibition of military to civil conversion. A row of minicars in front of the Agricultural Exhibition Hall caught my eye. They were small and crude in design, but General Zhang Aiping, the Minister of Defense, was eager to ride in one.

The weapons industry relied on imported technology to produce motorcycles, with good results. That generation of minicars was upgraded from motorcycles. They appealed to ordinary households. The engine displacement was from 450 to 550 cc, which was small. Interestingly, with the advocacy of Zou Jiahua, the Minister of the Ordnance Industry, most of the military industry's minicars used composite materials and weighed only 450–650 kg.

However, military industries were isolated and knew little about contemporary standards and trends. The rear seats in some models were just two benches facing each other, which seemed somewhat primitive. And where would the huge investment needed come from? Where was the market? And, since there was no way to obtain state approval, it was a dead end. On the eve of the Central Committee's decision to establish a car industry, a second trial minicar had failed.

2. The family car makes its appearance

My 1989: I hope it isn't a dream

Xinhua's *Outlook Weekly* published my article, "I Hope This Is Not A Dream," in its second issue of 1989. It was the first article published by an organ of the Central Government promoting family ownership of cars.

The author's article, "I Hope This Is Not A Dream," published in Xinhua's *Outlook Weekly* on 9 January 1989, advocating private cars for the first time.

My thinking was driven by economic considerations. I suggested that it was necessary and feasible to break through restrictions and permit families to have cars as soon as possible. I wrote:

"There's no harm in using reverse logic to break the deadlock. What would happen if we gradually privatize cars? The first to become wealthy have nothing to spend their money on. Like the workers, teachers, and public servants, their money goes to food, clothing, and expenses. With unbalanced income levels and overly-narrow consumer areas, it would be strange if the prices of non-staple foods and consumer goods don't go up! Why not open the floodgate and privatize cars? Commercializing housing is like canceling grain subsidies, as everybody is affected by it; but privatizing cars is like raising prices on name-brand cigarettes and wines, as only those who want are hooked. It isn't a threat to ordinary people's shopping baskets, but in fact it opens an overflow channel to stabilize prices."

"Buying a private car is not a single investment. Consumption continues in road maintenance, insurance, parking, and replacement parts. Only official vehicles travel on some new toll roads and the State pays twice, building the roads and buying the cars. We can only turn consumption into national construction by privatizing cars. 70 percent of road construction costs in Japan come from private car charges. Doesn't that tell us something?"

"More than 100,000 officials commute in publicly-owned vehicles. In fact, a foreign practice could be introduced. Apart from senior national leaders and veteran cadres, gradually privatize the vehicles of most of the officials and the heads of small and medium enterprises, with the State subsidizing the price of the car. Encourage leading cadres to buy private cars and drive them themselves. That would reduce consumption and improve relations between cadres and the masses. And it could stop the race for luxury cars as official vehicles. Younger leading cadres will welcome this approach."

State Councillor Zou Jiahua read the article and briefly commented on it: "Some people have already bought private cars, and a lot more want to buy them. The problem is there are no cars at suitable prices. A plan should be made to promote minicars, using the strengths of the defense industries. It would be best to go down the road of plastics or fiberglass."

Today we cannot help but feel the great changes in the Chinese economy. To be truthful, a private car was definitely an unreachable dream in those years. It was like what Momose said of Japan in 1950 — only one or two people out of ten thousand dared to dream of a family car. I put my dreams down on paper. I was also inspired by Ms. Song Huaigui, Pierre Cardin's representative in China. She said, "If someone asks me if I could pluck the stars from the sky, I wouldn't say no straight away. I would say, let's give it a try."

However, the political turmoil that happened soon after, and the ensuing debate between socialist and capitalist reform, flowed over into the economy. People even criticized housing reform as bourgeois liberalization. The danger was only averted when a speech by Deng Xiaoping in the early 1980s discussing housing reform was discovered. But in these circumstances, the idea of developing private cars and reducing the proportion of official vehicles was shelved.

The sound of knocking far and near

Economic restraint was relaxed at the end of 1991, and authorities began to consider minicars. Zhu Boshan left his enviable position as General Manager of China National Automobile Import and Export Corporation, going alone to the Guizhou Aviation Industry Base deep in the mountains to serve as the first General Manager of United Motors.

Although negotiations for the Subaru 550, registered as the Lark, were going on, I wrote:

"In the century-old history of cars, no country has ever been able to break away from two rules. One, only cars and trucks together make a complete car industry. It has taken China 30 years to understand that. Two, for the automobile industry to become a pillar industry, production has to reach 2 million, and there has to be mass consumption. The US reached that moment in 1920, West Germany in 1960, and Japan in 1966. When will China reach it?"

Deng Xiaoping made his southern tour the following year. Market economic reforms took a firmer direction, and enthusiasm for cars started up again. I went to Guangxi and other places with State Councillor Zou Jiahua to study the regional economy. He was interested in cars, and I spoke about the proliferation of official vehicles with him during the trip. He said, "People like to say that cars are a pillar industry, but that can't happen until families start having them."

Back in Beijing, I wrote "The Sound of Knocking Far and Near," a discussion on family cars. I gave it to two newspapers, but the editors didn't publish it because they were worried about my criticism of official cars, and about promoting high consumption. Only Ai Feng, the Director of the Economics Department of the *People's Daily* and a well-known journalist, would accept it. He even published it as a headline story.

Like a character in the Arabian Nights, I had experienced ups and downs for more than ten years, but I always had an unstoppable impulse. For a hundred years after the birth of the car, Americans, Europeans, Japanese, Koreans, and Brazilians enjoyed a car culture, expanded their scope of travel, and created huge social and material wealth. Why couldn't Chinese people do that?

3. Cars are bright and beautiful

The family car is written into industrial policy

The China National Automotive Federation was reorganized as the China Association of Automobile Manufacturers in 1990. Its industry management functions were incorporated into the newly established Automotive Department of the Ministry of Machinery Industry in October 1993. Lu Fuyuan was the Vice Minister, Zhang Xiaoyu was Director of the Automotive Department, and Miao Wei was Deputy Director. Since Reform, the Automotive Department was the third and final functional management organization of the automobile industry after the China National Automotive Industry Corporation and the China National Automotive Federation.

In the 1990s, people understood more clearly that family car ownership would support economic growth. Zhang Xiaoyu was a firm supporter. He told me that in an intergovernmental exchange between China and Korea, his counterpart asked, "When is China going to start developing its car industry?" He answered, "We are still solving the problem of having enough to eat." But the other man said very earnestly, "It was precisely when people still didn't have enough to eat that Korea tightened its belt and started developing cars. If we only relied on farming for food, most Koreans would still be getting by like our relatives in the north, who can't even drive a car in their dreams." Zhang found that enlightening. I put the conversation on the cover of my book *The Lure of Family Owned Cars in China*.

I visited Hyundai Motor Co., Ltd. for the first time in summer 1993. I entered the city from the airport by the elevated road along the Han River. Almost all the cars zooming each way in the six lanes were Korean. Korea put only a 10 percent import tax on cars, but China's 180–220 percent tariff barriers couldn't stop the flood of imported cars. This wasn't explained by the phrase "Korean patriotism." On the streets of Seoul, I found that watermelons were five times more expensive than in China, while the price of a car was only one-fifth as much. I saw endless rows of cars at Ulsan Wharf being loaded onto a 100,000-ton freighter. Three million Hyundai cars were being shipped to the US every year.

The new generation of Chinese policy makers was impressed by the way Japan and South Korea developed their automobile industries and grew their economies in the 1960s and 1970s — by relying on government policy support and working stubbornly with businesses.

The Chinese Government announced the Industrial Policy for the Automotive Industry in April 1994. For two years before the announcement, I participated in many meetings chaired by Deputy Prime Ministers Li Lanqing and then Zou Jiahua to formulate and revise the draft policy. From guiding ideas to final document, every word of the text was also scrutinized by the Central Financial and Economic Leading Group and the highest decision-making levels of the State Council.

The legality of purchasing private cars in China was officially recognized for the first time in the policy. Although it avoided direct mention of family car ownership, it was still a breakthrough in the decades-long restrictions on private cars.

These are some of the articles in the policy relating to family cars: Article 1: Total automobile production will meet the needs of more than 90 percent of the domestic market by 2000. Production of cars will reach more than half of total vehicle production, and basically meet the needs of family ownership.

Article 46: The consumption structure of purchasing cars with public funds for use mainly by administrative organs, organizations, institutions, and state-owned enterprises will gradually change.

Article 47: The State encourages individual purchase of automobiles and will formulate specific policies at appropriate times, in accordance with the development of the automobile industry and changes in market consumption structure.

Article 48: No locality or department shall interfere by administrative or economic means with the purchase and use of vehicles of legitimate origin. Active measures shall be taken to support and guarantee facilities and systems such as license management, parking areas, gas stations, and driver training schools.

Article 54: As more people own cars in each area...the reconstruction and expansion of urban roads shall be an important task of urban planning and shall be implemented promptly.

Article 55: Starting from the 1995 school year, primary schools shall increase traffic awareness by including traffic knowledge in their teaching content.

The right of ordinary people to own cars wasn't recognized at China's highest policy-making level until 45 years after a socialist China was established. However, it was hard to break old habits. As the articles in the policy had too many basic principles, the detailed provisions were never implemented, and there were too many separate jurisdictions with departmental and local interests at cross purposes. Plus, grassroots officials directly opposed family ownership. Private cars became a dead letter in the Industrial Policy.

Cars were still something that everyone wanted but nobody could get. In some provinces and cities, in addition to paying considerable national taxes and fees, car buyers were charged for more than 20 payments, like car purchase adjustment funds, social control fees, fixed fees, urban expansion fees, road construction funds, license fees, and urban education funds. One city in Jiangsu that wasn't even on the rail line actually imposed a railway crossing fee. Those taxes and fees were sometimes 1.2 times as much as the price of the car.

It became chaotic. Villages and towns all along the roads set up roadblocks and collected fees for traffic, public security, industry and commerce, forestry, and environmental protection. Farmers even dug a pit in the road, laid wooden boards over it, and charged cars to go across.

International giants attend the 1994 Family Car Forum

In contrast with the stagnation the Industrial Policy for the Automotive Industry ran into domestically, it was the major multinational automobile companies which responded to it. Taking advantage of the policy, the CEOs of almost all the well-known automobile manufacturers flocked to China to explore possible cooperation.

The 1994 Beijing International Family Car Forum was held at the China World Trade Center at the end of that year.

The government announced that in the next two years, it would concentrate on four major production bases — FAW-Volkswagen, Shenlong, Shanghai Volkswagen and Tianjin Xiali, each with annual production of 150,000 vehicles. No new car projects would be approved until the end of 1996, but China welcomed international cooperation in car components. The approval of new car OEMs in the future would depend on growth of the Chinese market and industry. Foreign manufacturers which had cooperated with China's auto industry in components during the past two years would be given priority.

These vague words caused turmoil among the "knights in shining armor" in the global car industry, causing them to fight for the prize of the Chinese family car.

The host of the Forum reassured global manufacturers. In a report on behalf of the Chinese government, Miao Wei, deputy director of the Automotive Department of the Ministry of Machinery Industry, said, "The Chinese government will further encourage the automobile industry to cooperate with foreign countries and encourage businesses to make use of foreign capital and technology. When Chinese companies choose their partners, they will first choose those which have product patents, product development technologies, production management, manufacturing technologies, independent sales channels, and sufficient financing capabilities. Foreign companies can expand joint ventures in existing car projects and jointly develop products."

The international car industry tried every which way to enter China's family car market.

A few months after the policy was released, Mercedes-Benz developed its new Family Car China (FCC) model and produced prototypes for display at the Forum. At its booth, Porsche sought favor with Chinese families by exhibiting the newly developed C88.

Almost all multinational automakers entered the fray, sparing no expense in research and development aimed at the Chinese market. They weren't looking at the hundreds of thousands of official cars, but at the potential demand from millions of ordinary families. Their long-term plan was naturally to gain a position in the last potential large market in the world.

The Forum was only a gesture to the multinationals. No company was given the go-ahead to produce family cars. However, the auto show that accompanied it caused a sensation in Beijing. After that there were car fans in Beijing and China for the first time.

FCC drives onto the racetrack

I was among the Chinese journalists who visited Mercedes-Benz in June 1995.

At the test site in Unterturkheim, Stuttgart, an FCC was parked beside the circular track. In contrast with the huge test track, it seemed much smaller than when I saw it in Beijing.

I was the only journalist with a driver's license. Mr. Ellison, the head of the FCC development team, handed me the keys and asked me to try it.

It was similar in size to the Xiali. I opened the door and got in. The driver's seat was roomy. I turned to look at *People's Daily* journalist Cao

Huanrong who was in the back seat. He said it was comfortable and there was plenty of space between his knees and the front seat.

I buckled the seat belt and drove onto the track. The steering was very light, and the 1.3-liter engine was powerful. I easily overtook two big cars doing routine tests on the track. Although the body was small, it didn't drift at more than 120 kilometers per hour. When I raced into the corner on the second lap, I didn't slow down because I wanted to see how driving on the 45-degree raked track on the outside of the curve felt like. Sure enough the horizon outside the window tilted up, but with the car's superior performance at high speed I didn't feel anything. However, my colleague Cao, who was in the back seat, said with some feeling that he hardly ever perspired, but his palms were sweating when he saw the concrete retaining wall whizz by.

I drove off the track and everyone gathered around. Mr. Ellison gave a thumb-up: "You've driven Mercedes-Benz's most expensive car." I learned that when developing a new car, Mercedes-Benz usually made about a dozen for testing. However, they only made one FCC prototype.

The FCC was born out of the A-series sedan that Mercedes would launch in 1997. Following the sandwich design principle, the floor of its body had two layers. The engine, gearbox, suspension, and fuel tank were housed in the lower space, making the upper space comparable to a midsize sedan. This design also made the FCC very safe, with a clever design that slid the engine and gearbox under the passenger compartment when the car was in a head-on collision. In order to suit Chinese income levels, the price was kept under US$10,000, but improvements were still made to the revolutionary FCC.

Who said Germans are conservative and old-fashioned? When a potentially huge new market emerged, they responded decisively.

Without the Chinese government's approval, Mercedes-Benz couldn't find a company to cooperate with in producing its family car. I have visited Mercedes many times since, but I have never seen the lovely FCC anywhere in its museums and design centers.

The FCC developed for China originated from the Mercedes-Benz A-Class "sandwich" design.

The author and *People's Daily* reporter Cao Huanrong posed for a photo after test driving the Mercedes-Benz FCC.

A fruit you can only pick on tiptoe

Small cars began to attract my attention.

In Europe, people favored small cars because of the high cost of gasoline and because cars were purely a means of travel. Sales of small cars were 4.5 million a year in Europe in the 1990s, making up 35 percent of total sales.

I counted cars driving by in the streets of Paris and found the proportion of hatchbacks was as high as 60 percent. Small cars were the fashion in the big city.

Ford introduced the Ka, a newly developed compact car at the Paris Motor Show in October 1996. It was midway between the domestic Xiali and the Alto.

To take the competitive lead, Ford used the largest computer systems in the global business community at that time, turning over traditional drawing, modeling, and testing processes to the computer. I saw the most advanced testing equipment and computer-aided design systems, with a total investment of £3 billion, when I visited Ford's Denton Technology Center in London, where the Ka was developed.

The production line left its mark on the first century of automotive industry. Its next century would see a mix of emerging technologies: computers, global satellite positioning, robust lightweight materials, new fuels, and perhaps most importantly, electronics. I heard something interesting at Ford. They were saying cars were computers on wheels. The content of electronic devices in cars was basically zero 20 years ago. They made up about 10 percent of the cost in the 1990s. That could be expected to double in the near future.

They told me that before the Ka went into production, Ford organized three test teams: the Ford driving group, which consisted of design experts; a professional drivers group, including Formula One drivers; and a women's group, housewives who were employed to counteract the tendency in car design to favor men. This showed that car design was being humanized.

A vague image of a Chinese family car was becoming clearer. I felt that the simple models for a family of three that were coming out domestically were far from international standards. What would a

Chinese family car look like in the 21st century? I wrote in 1995 that a family car was a fruit you could only pick standing on tiptoe.

I didn't think the Chinese family car should cut corners by just fixing a shell onto a motorcycle. The technical standards had to keep pace with international models of the same class: to carry a family of five with three generations comfortably. Considering its speed, air conditioning, and audio needs, the engine displacement should be around 1.3 liters. Family cars are related to ordinary people's lives. We had to give top priority to reliability and safety and ensure that there was no need for overhaul within 100,000 kilometers. These conditions seemed strict at the time, but they were as reasonable as for color TVs and refrigerators that could be used for three to five years without maintenance. One of the preconditions for family use of color TVs was the thorough assurance of quality. The quality of family cars should be higher than that of color TVs. Cars are the most expensive consumer items, and they directly support people's lives. If one out of 10,000 cars has a quality problem, it's enough to bring a company down. Therefore, blindly keeping down costs without regard for quality and safety shouldn't be the goal of the Chinese family car.

A Chinese family car was a fruit that was almost out of reach. That was recognized by many people in the industry. However, the choice Chinese people made was the bigger the better. The only similar trend in Asia was in South Korea. The economic explosion came too swiftly, and it was dictated by the mentality of the newly rich.

4. The long narrow road

A battle of words with the Deputy Premier: A step too high

In the mid-1990s, the news media frequently reported on residential construction as a new point of economic growth. To end allocation of low-grade housing, housing reform based on gradually commoditizing housing progressed rapidly in cities and towns across the country, spurring the desire of many Chinese for home ownership.

The demand for housing led people to travel further. Private cars could have promoted its commoditization. However, calls for family cars were received coolly because they were misunderstood as rivals for people's limited house-purchase funds.

The auto industry held a conference at the Great Hall of the People in July 1996 to mark the 40th anniversary of the birth of the Liberation truck. When the papers were brought together in a volume, mine was included.

Machinery Industry Minister He Guangyuan wrote to the Central Committee leadership in October. He talked about indigenous research and development of Chinese cars and family car ownership. He wrote at the end of the letter, "Authorities have been discussing social consumption. They differ on whether to solve housing or vehicle transport first. Xinhua journalist Li Anding presented a paper at the conference which might inspire the leadership's policy-making. Attached for your reference."

When I first met the Minister in the mid-1980s, he was one of the youngest ministerial cadres in China. His views were sharp and logical. We became friends despite our age difference. I listened to his advice for many years, and we often found ourselves in agreement.

Minister He joined the army during the War of Liberation. He specialized in metal punch pressing at Kiev Institute of Technology in the USSR in 1952. He went to FAW and became director of the forging factory in 1956. After the Cultural Revolution, he served as Director of Changchun Tractor Factory. In 1980, at the age of fifty, he became Deputy Director of the Agricultural Machinery Department. He was the Machinery Industry Minister in the 1990s. He became a leader in the auto industry and a promoter of family car ownership.

Minister He attached my article, "The Space Brought by Wheels," in which I wrote:

"Transport and housing are opposite but inseparable entities, and they are two adjacent rungs on the consumption ladder. Stepping off into thin air is unavoidable if a rung is missing. A step of more than 10,000 yuan for transport would be punished economically."

"If we make housing commercialization a strategy to improve the structure of housing consumption, promotion of family cars doesn't have to be a rival. Policymakers must be made aware that the privatization of cars is a driving force for the economy and a catalyst for the commercialization of housing."

With an economical family car, people's travel radius will expand. Instead of a million yuan house in the city center, they can seek a more affordable 100,000 yuan residence in the suburbs. Residential areas in new urban zones will be transformed. If we look at it that way, how can family cars cut into housing construction?"

On 15 October, General Secretary Jiang Zemin added a comment to He Guangyuan's thinking on the letter:

"Many problems in the process of economic reform still baffle us today. Finding a better balance between foreign investment and development of national industries is one. Finding solutions to problems such as choice of car model, pricing, environmental protection, parking areas, and roads is another. Those are big industrial policy issues and the State Council should study them seriously."

"Some people are not opposed to family car ownership, but ask how it can happen and how fast. With the current widespread economic growth, doing things on a small scale and with low quality has caused overstocking. It could also destroy our hopes of cars being a key industry."

My article was quite long, but Deputy Premier Zhu Rongji apparently read it through carefully. And although he didn't agree with me, he made this brief comment:

"The Xinhua article has literary style but lacks macro understanding. We have insufficient oil resources and we have become a net importing country. The streets of our cities are crowded, people's purchasing power is limited, and people still have difficulty getting a house, let alone building a garage. With all this, how can we put cars first, and rely heavily on them to solve the transport problem!"

According to instructions from Jiang Zemin, Zhu Rongji and other leaders and led by the State Planning Commission, more than a dozen

最近，我听说有关部门正在议论引导社会消费的问题，对先解决"住"还是先解决"行"看法不一。就这个问题，新华社记者李安定同志在这次座谈会上有个书面发言，或许对领导决策有所启发，随信附上，敬请参阅。

一九九六年九月五日

3

Deputy Premier Zhu Rongji's commenting on a letter from He Guangyuan about the author's article.

ministries and commissions began putting forward their particular views, and began a comprehensive policy study in April 1997 with a view to development into the new century. That became the Research Report on Several Major Policy Issues in the Development of China's Automobile Industry.

Over the next two years, except for discounted sale to residents of previously allocated housing, housing sales were slow and vacancy rates in the new residential areas remained high. The economy made a soft landing but urgently needed a new point of growth. Although family car ownership had been suggested by experts and authorities many times, even recommended by the State Planning Commission, it was blocked again and again, in adherence with the established thinking of the State Council leadership. It never became that point of growth.

It wasn't until the Tenth Five-Year Plan that the expression "encourage family car ownership" was included.

2001: A family car is a right

In March 1998, the Writers Publishing House launched my major work entitled *The Lure of Family Owned Cars in China*. It became a bestseller. Chen Jiangong, a well-known writer and Party Committee Secretary of the Chinese Writers Association, said, "The significance of this book goes far beyond the scope of literature. The ideas it puts forward will leave their mark on Chinese society."

Leaders and friends from the Department of Textiles, Light Industry, Electronics and Machinery of the State Planning Commission, the Ministry of Machinery Industry and its Automobile Department, as well as the media, attended a seminar on the book. Media friends were surprised that Chinese auto industry authorities were there. It's interesting that as we look back at the history of Chinese automobiles, and especially at the car industry, "tragedy" is one of the most widely used words.

The road was long and narrow, but I followed it to the end as the wheels of time continued to roll forwards. I refined my understanding of family car ownership in that ten years. From considering it an economic breakthrough yet to be made, I now thought of it as a basic right of ordinary people in a "people's country."

Xinhua started my column "Car Talk for Laymen" in March 2001. In the first essay, I wrote:

> More and more Chinese dream about owning a car. What is a family car to them? Discussion bubbles on in the media. Should it be a sedan or a hatchback? Should it cost 100,000 yuan, or 80,000? First of all, we have to be clear what a family car is.
>
> It's a right. The enjoyment of car culture is a right modern people should have, especially in a Socialist society. It can't be blocked forever by policy restrictions, nor will it be bestowed as a good action by wise leaders. Government should take effective measures to develop the automobile industry, improve the user environment, and formulate strict environmental protections and safety regulations, to ensure people's basic right to pursue a higher quality of life.
>
> Approval of family car ownership reflects progress for ordinary people in China from having duties to having rights.

Chapter 4

Mutual Profit is the Important Thing

I t was the eve of the 50th anniversary of the founding of New China on 6 September 1999. In Changchun, FAW-Volkswagen started its production line for the Audi A6, representing a world-class, mid- to high-end sedan.

It's fresh in my memory because, to create awareness of its new luxury brand, FAW-Volkswagen invited guests and the media to fly to Changchun in a Boeing 737 aircraft. It also provided an electronic version of the press release and internet service for the first time in China. This was new to most of the guests.

The concept of quality attached to the slogan "One planet, One Audi" showed FAW-Volkswagen had reached the same standard of high-end automotive technology as Germany.

China was becoming more open at the beginning of the new century. Themes were changing from war and revolution to cooperation and peaceful development. Its auto industry had struggled with inferiority, but it was moving toward mutual profitability and equality with the international auto industry.

Mutual profitability had become the keynote of a number of new joint ventures in the new century. The Buick Century, Honda Accord, and Audi A6 sedans were all coming off the assembly line. The Chinese auto industry had a new group of mid- and high-class sedans.

1. Passat: A secret affair

Demand presents another challenge

The government had granted special approval for FAW to import and assemble parts for 890 Mercedes-Benz 280 sedans in the late 1980s to cope with demand for upgraded official vehicles. They were distributed to ministries, provinces, cities, and large state-owned enterprises. Their quality wasn't ideal and they disappeared from the streets in a few years.

After several years of economic contraction caused by administrative restructuring, Deng Xiaoping delivered his Southern Tour speech in 1992. The speech proved inspirational and the economy surged. Within a year, there were four times as many new companies established as in previous years. Demand for mid- and high-end cars soared and their prices rose.

The Santana, the Jetta, and the Fukang hatchback could no longer satisfy official and private demand. Imports like the Toyota Crown and the Nissan Duke (called the Cedric in Japan) passed through the hands of speculators, reaching astronomical prices.

Jiang Zemin looking at a Santana 2000 taxi at the Beijing International Auto Show in 1994, as the author (third from the left in the front row) showed him the satellite navigation system.

The supply and demand imbalance made cars the most highly sought after smuggled goods. Car smuggling ran rampant again.

Local customs seized 6,791 smuggled cars worth 1.53 billion yuan in the first half of 1993. This was 4.7 times as many as in the previous year.

The lawbreakers who lusted after the huge profits weren't just from the criminal ranks, but included some legal enterprises and institutions. There was a huge underground market and domestic consumption gobbled up hundreds of thousands of smuggled cars. Prices also ratcheted up as they passed from hand to hand.

Smuggling cars was difficult and risky, and not many foreign gangs were interested. So, why were so many people in China attracted to it?

Cars are undoubtedly among the highest priced industrial products in the world. However, car prices are excessive in China due to man-made factors and because of the huge gap between market supply and demand.

Thanks to the benefits of Reform, many Chinese have gone abroad and have seen the outside world. A common experience they all have is that products are more expensive outside China, but cars are cheaper, even much cheaper than at home.

A friend who lives overseas summarized it in this way: in developed countries, goods that are eaten, worn, and used are generally three to five times as much as similar goods in China, while the price of a car is just one-third as much as in China.

The import tariff on cars of less than 1.3-liter engine displacement was 180 percent in 1993; for cars of 1.3–3 liters, it was 200 percent; and for those with a displacement of more than 3 liters, it was 220 percent. A car tripled in value as soon as it passed through customs, and then there were vehicle purchase fees, special consumption taxes, and sales profits added at each level. The price of one car in China would buy at least four abroad.

Naturally, high tariff barriers were intended to block the dumping of foreign cars and protect the national automobile industry. But the heavy taxes curbed consumption and efforts to grow the domestic car industry weren't getting anywhere. People eager to get rich overnight couldn't resist the lopsided prices.

The average tariff on imported cars was about 15 percent in developing countries, and 5 percent in developed countries. Taiwan also imposed high tariffs. However, as its car industry grew, they were lowered by 6 percent a year. By 1991, they had fallen from 65 percent to 30 percent. This anti-protectionist policy enabled Taiwan's automobile production to double in five years. But mainland China's high import tariffs remained and were the highest in the world.

Adam Smith wrote in *The Wealth of Nations*, "A heavy tax sometimes reduces the consumption of goods that are taxed, and sometimes rewards smuggling." The Chinese car market is an appropriate footnote to the conclusions drawn by the British economist a hundred years ago.

Chinese aesthetic standards for automobiles were also formed in the 1990s. They came from the Japanese cars that filled the streets. The Toyota Crown and the Nissan Duke were boxy, standard sedans. FAW-Volkswagen chose to introduce the popular Golf hatchback and the Jetta sedan, both of which were on the same chassis. Before production began, the two models made a run of 30,000 kilometers from the snowstorms in the north to the coconut groves on Hainan Island. Eighty percent of people who saw them along the way said the Jetta was the best and most stylish.

The Tianjin Xiali minicar had just gone into production. This technologically new Daihatsu hatchback from Japan was trendy and had a round shape, representing the fashion in small cars globally. However, in China it was not well received, with some people calling it the Shoe. This was a disappointment for Tianjin Xiali, so they developed a sedan in 1992. People said that adding a rear compartment ruined the aerodynamic performance and compactness of the original model, and increased the cost by twenty or thirty thousand yuan. The response to the market survey was unanimous. It didn't matter that it cost more, passengers were willing to sit in the back, and drivers loved it.

Fan Jingyi, Editor-in-Chief of the *Economic Daily*, started a special weekend edition edited by Li Dongdong. I wrote an article for the inaugural issue, "How I Feel About the Lengthened Xiali." I wrote that for Chinese to call a car a sedan, it had to be high in the middle and low at both ends, shaped like an ancient sedan chair. This perception was more important than the car's speed and handling, and it affected production. When people got married, they chose a sedan, just as officials did. The hatchback had no back seat, so no one in China would buy them, even though they made up more than 60 percent of cars in European cities, and they looked smart and fashionable. What would satisfy the market: catering to old tastes or scientifically developing new opportunities? You shouldn't underestimate how ready Chinese can be to accept new things.

Another criterion was the more knee space in the rear seat, the better. I called it the crossed-leg coefficient. At the turn of the century, modifications had to be made to bring in executive cars. There were problems with road conditions, oil quality, differences in temperature and humidity between the east and the west, but none of these was as important as the crossed-leg coefficient.

After more than ten years of interaction, foreigners finally realized that the major feature of models imported to China was making them longer.

Playing games and compromising

People criticized Volkswagen for building the Santana for more than a decade without developing any new products, and making as much money as possible with an obsolete model. Very few people realized that model selection or when to introduce them wasn't something car companies, or even joint ventures, could decide for themselves.

At the beginning of the 1990s, Shanghai Volkswagen responded to stronger market demand for mid- to high-end cars. It informed the Shanghai authorities that it would bring in the latest generation of B-class vehicles, the Passat B5, and would produce it in China as early as 1995. The application was held up in Shanghai for a while. When it finally reached Beijing, authorities there were slow to respond.

Dr. Li Wenbo, who twice served as the chief representative of Volkswagen in China, recalled, "When I came to China to take over in 1996, I had been urged repeatedly by Volkswagen Asia-Pacific Regional Manager Martin Posth to find out why there was no response to the

Passat B5 application. Posth didn't understand what a Chinese problem it was."

Li Wenbo finally figured out that Passat was stuck for two reasons. Firstly, the Central Government had strict controls on big projects like new cars. Shanghai Volkswagen already had the Santana 2000, and approving another new model would unfairly disadvantage other car manufacturers. The State Economic and Trade Commission couldn't deal with it. Secondly, the Shanghai government was trying to promote joint venture projects with General Motors (GM). The Passat project could only take shape when the GM project was approved.

Ferdinand Piëch[1], who had replaced Hahn as the chairman of the Volkswagen Group, visited China in 1997. He met Vice Premier Zhu Rongji in the Purple Pavilion at Zhongnanhai, and requested approval for the Passat project. Zhu Rongji told him, "You should focus on the Santana. You've done a good job with the Santana 2000. Why not develop a Santana 3000?" Zhu Rongji flatly turned down Shanghai's new model and there was no room for maneuver.

The allure of the mid- to high-end car market was too great. Moreover, after ten years of development, Shanghai Volkswagen had produced 500,000 Santanas. The product mix was in urgent need of an upgrade.

Shanghai people were quick-witted. Shanghai Volkswagen introduced the Passat under the guise of a technical improvement project on the Santana. The Santana was VW's B2 (the second generation of the B-class car) and Passat was the B5, so it could justifiably be called a technical improvement. Also, Shanghai Volkswagen's Third Plant,

---

[1] Ferdinand Karl Piëch (born 17 April 1937) is an Austrian business magnate, engineer, and executive who was the chairman of the executive board of Volkswagen Group in 1993–2002 and the chairman of the supervisory board of Volkswagen Group in 2002–2015. A grandson of Ferdinand Porsche, Piëch started his career at Porsche, before leaving for Audi after an agreement that no member of the Porsche or Piëch families should be involved in the day-to-day operations of the Porsche company.

specifically for production of the Passat, was already under construction. However, perceptive personnel at the State Economic and Trade Commission said, "Don't take us for fools. How is this an improvement on the Santana? You're just introducing a new model." But there was no turning back. The project went forward after a lot of coordination with the Shanghai Municipal Government and other departments.

There had to be some changes made to the prototype since it was a technical upgrade. However, Piëch, who was particularly wedded to technology, thought the Passat B5 was fine as it was. Changing one aspect would affect the overall technical parameters. The deputy mayor of Shanghai told Piëch that the Passat wouldn't be approved unless it was changed. But he didn't say how. Piëch gave in and said, "We can change it, but we won't lower its performance." The Shanghai Volkswagen Lengthened Edition Passat, with a 100 mm longer wheelbase, came off the production line two years later.

The first time Zhu Rongji met with representatives of Shanghai Volkswagen after the Passat came on the market, he said indulgently, "You have made small changes in this car. Now it is a fait accompli, so we'll have to turn a blind eye." And so, production and sales of the quasi-legal Passat began under the guise of the third generation Santana (PASSAT). But the Passat didn't gain a foothold in the domestic mid- to high-end sedan market until 2004, when it had a major upgrade and a price adjustment.

At the Passat launch, Shanghai Volkswagen General Manager Hong Jimin called Passat the third generation Santana (PASSAT). A reporter who asked why it was called that rather than just the Passat, he was told that it was sensitive and couldn't be put in writing.

The Passat's introduction was a real drama. Central Government authorities were following the rules, while the Shanghai government and VW Germany each had their own agenda. The three eventually reached a compromise. Reform was a process of constantly breaking old rules and making new ones.

After the economic volatility of the 1990s, the car industry gradually matured and gained confidence. It no longer needed to feel inferior.

After ten years, the Chinese sedan was ready to take off. In addition to Shanghai Volkswagen's low-key testing of the water with the Passat, factories with the most advanced equipment in the world were built at Shanghai GM, FAW Audi, and Guangzhou Honda for three new mid- to high-end projects. This showed a change in thinking. We bid farewell to ideological struggle and isolation, and moved toward mutually profitable cooperation with the international automobile industry.

Many foreign companies had given their Chinese partners old products that had been on the market for many years. From December 1998 to the eve of the 50th anniversary of the founding of New China in 1999, the new Buick Century, the Honda Accord, and the Audi A6 had been rolling off their production lines, forcing international rivals to compete with advanced products. A new group of medium and high-class cars began to appear.

2. Discussion and negotiation: From Audi 100 to Audi A6

Why did Audi go to FAW Volkswagen?

Volkswagen entered into the Audi pilot project with an annual output of 30,000 vehicles in order to get a joint venture project with FAW with an annual output of 150,000 vehicles, while FAW was simply introducing technology to produce a second generation of Red Flag sedans with accumulated funds.

FAW got the technology for the Audi 100 and built welding, painting, and final assembly lines at the old Red Flag site. Second-hand equipment for the production of body panels was purchased from the Audi factory in South Africa. The cars were put together from SKD and CKD kits while the factory was being built.

The first Audi 100 sedan came off the assembly line on 1 August 1989. A total of 1,922 vehicles were assembled that year.

Audi's predecessor, Horch[2], existed for 100 years. Its founder August Horch[3] always went for technology and luxury. The price of a Horch 853 sedan was 14,900 Deutsche Marks in 1930, equal to the price of a house. Audi, Horch, Wanderer, and DKW came together to form Auto Union in June 1932. The company's badge was the four rings that people know today. Auto Union's East German factory was looted by the Soviets at the end of World War II. The employees of Auto Union regrouped in the small town of Ingolstadt, near Munich in West Germany to restart the business at a spare parts warehouse.

Auto Union was later acquired by Mercedes-Benz and Volkswagen. It developed its first new model in 1968, keeping the four-ring badge. This was the Audi 100, which revived the Audi brand. The third-generation Audi 100 (C3) was named World's Best Car of the Year in 1984. Audi sparked interest with its low drag coefficient of 0.3, lightweight body, and quattro full-time four-wheel drive technology.

In 1987, I rode in the car of the Secretary General of the China National Automotive Federation, Chen Zutao, and asked about its four-ring badge. Chen proudly said, "The Audi 100 is as advanced as a Mercedes-Benz." It must have been one of the Audi 100 assembled by Shanghai Volkswagen in the early years.

FAW produced the Audi 100 under a 1988 license agreement with Volkswagen using CKD kits for six years. Audi sent technical and management personnel to assist with quality control and localization, progressively reducing the proportion of imported parts.

---

[2] Horch was a car brand manufactured in Germany by August Horch & Cie, at the beginning of the 20th century. It is the direct ancestor of the present day Audi company, which in turn came out of Auto Union, formed in 1932 when Horch merged with DKW, Wanderer and the historic Audi enterprise which August Horch founded in 1910.

[3] August Horch (12 October 1868–3 February 1951) was a German engineer and automobile pioneer, the founder of the manufacturing giant which would eventually become Audi.

FAW brought in Audi 100 technology to produce a small Red Flag sedan in the early 1990s.

The Audi 100 completed its mission in China six years later. FAW changed the badge from Audi to Red Flag when the localization rate reached 82 percent in 1995.

The Audi 100 revitalized FAW. It produced 100,000 Audi 100 cars and then 23,000 small Red Flags between 1988 and 1997. The localization rate reached 82 percent and then 93 percent. Revenue was 31.1 billion yuan in the 10 years of sales. 6.6 billion yuan in profits and taxes was turned over to the government. This was 10 times the 625 million yuan investment.

Though it had good talent, management, equipment, processes, and experience, FAW, which had only produced trucks for many years, was far from being able to produce high-end cars. Quality control of localized parts was constrained by domestic machining standards and electrical equipment manufacturing, which could not be fixed quickly.

The biggest problem was that Audi could not take part in management or be in control. Senior Audi executives, seeing how FAW produced the Audi 100, bluntly called it a mess and nearly wanted to quit. Quality standards weren't being met, failure rates were high, and its reputation was eroding when the technology transfer agreement was about to expire, but Audi was reluctant to give up the potential of the China market.

Volkswagen Chairman Piëch, Audi CEO Herbert Demel, and FAW Factory Director Geng Zhaojie discussed it in detail. Geng and Demel often flew to Frankfurt for talks. They eventually decided to take Audi production into the FAW-Volkswagen joint venture.

FAW, Volkswagen, and Audi signed an agreement in Beijing on 13 November 1995 during a visit by German Chancellor Helmut Kohl. Volkswagen gave 10 percent of its 40 percent shareholding to Audi when FAW refused to give up any of its shares. The share ratio was changed on 18 December to 60 percent for FAW, 30 percent for Volkswagen, and 10 percent for Audi. FAW-Volkswagen officially began producing Audi products.

Fighting for a voice with persistence and compromise

Product selection was the first thing Audi had to do after joining FAW-Volkswagen.

In Germany, the Audi 100 was renamed the Audi A6 in 1994. Audi was preparing to launch a full-scale assault on the Mercedes-Benz E-Class and the BMW 5 Series. The greatly upgraded Audi A6 (C5), the fifth-generation C-Class, carried the burden of an intensively publicized research and development effort.

Since the C4 produced in Germany was already halfway through its life cycle, it wasn't worth the huge investment to bring it in. And although the C5 wouldn't be available until 1999, the FAW-Volkswagen shareholders decided to produce it at the same time as in Germany, despite the huge challenges.

For the next four years, FAW-Volkswagen kept producing the Audi 100 (C3), which with partial upgrades and a new V6 engine, was called the Audi 200. It was the only domestic high-end executive vehicle.

Cultural differences often caused friction between the joint venture parties, but they faced up to differences, each drawing from the other's strengths.

Lu Linkui, the second General Manager of FAW-Volkswagen, recalled, "When we were still feeling our way forward, a budget was often approved one day and revised the next. Volkswagen needed a budget. And when it was set, they considered it to be legally enforceable. The Germans paid attention to processes. Their specialists delved into technology, production, sales, and service. This increased the probability of success, which was difficult to achieve without it."

Having a voice had always been a sensitive issue in Chinese joint ventures and was the most likely to be politicized and nationalized. It was generally believed that the foreign party had the upper hand because it provided the products, technologies, and brand that the company needed, and the Chinese lost their right to speak.

Lu Linkui said the issue of having a voice should be resolved at the board level. Management personnel should be accountable to the board of directors, whether they were Chinese or foreign. However, in reality, Chinese and foreign management personnel each had a bias toward their own side and perceived the other to be selfish. They couldn't help it. How well they could bridge the gap often determined the success or failure of a joint venture.

The Germans were straightforward and pragmatic. Hermann Starke, who was in charge of Audi's project in China, said to me, "There were more than 10,000 Chinese working at FAW-Volkswagen in Changchun, and only about 50 European managers. Given this imbalance, we were focused on partnership. We didn't want the elephants to trample the mice."

Starke said, "Audi Board member Erich Schmitt[4] came to China often. It was important for us both to be on site. We tried to accept their ideas and grow with them. Our strategy wasn't just pulled out of thin air."

There had to be unanimous agreement at FAW-Volkswagen before major undertakings began. The keys to successfully running this uniquely structured, equity owned organization were respect, com-

---

[4] Audi Board Member for Purchasing Erich Schmitt was responsible for the Chinese market at this time.

munications, and compromise. The major shareholder didn't use its controlling position to infringe on the interests of the minority shareholder, and the minority shareholder didn't use its advantages in products, technologies, and brands to pressure the majority shareholder. Governance was harmonious. Shareholders and the country both benefited.

10 percent of the shares, 100 percent of the investment

You had to admire Audi, which only held 10 percent of the joint venture company. It was just beginning to grow again in the global market, and its brand image was still weak in 1995. It had to do a lot to catch up with Mercedes-Benz and BMW. With even the slightest misstep, it would all go down the drain. It was risky to rely on the A6 (C5) for its rejuvenation, given the technological level of FAW-Volkswagen.

Unlike with the Audi 100, as a shareholder Audi would be fully involved in the production of the A6 in China. It would be engaged in production management, parts procurement, marketing, and after-sales service of the Chinese-built Audi, and would bear the corresponding risks.

The agreement was signed in January 1996. The Chinese side requested simultaneous development of a lengthened version for the Chinese market. This prototype would be synchronized with development of the original Audi A6 (C5) at Audi's headquarters in Germany.

The Chinese sent six people, and Audi had more than 300 people working on the project over the next three years. R&D costs were paid by FAW-Volkswagen, and it owned the intellectual property of the lengthened model.

The simultaneous development of A6 (C5) satisfied the need for more rear passenger space for Chinese customers. When Starke took charge in China, he realized the Chinese wanted the lengthened version. German engineers hadn't picked up on the signals from the Chinese. Europeans and Americans felt it already had enough space. Why didn't the Chinese just accept it as is? They found that China was very different from Germany, where 90 percent of executives bought and drove their own cars. In China, 90 percent of them sit in the back seat. The Audi Board of Directors agreed to honor the Chinese request.

From the beginning, Audi took into account the performance, safety, aesthetics, beauty, and other aspects related to its lengthening. Even the trunk lid was specially designed for the Chinese version. "We didn't just cut it in half to lengthen it like other manufacturers did," Starke said.

Audi had learned a painful lesson with the Audi 100 in China. As it prepared to start production of the Audi A6 (C5), it wouldn't make exceptions and insisted on having the final say on quality. Aircraft from Beijing to Changchun were filled with Germans from Ingolstadt. If you walked into the workshop where the Audi was being assembled, it was like going to Germany. Hundreds of fair-haired Audi management personnel, engineers, and technicians worked side-by-side with the Chinese workforce, passing on their technical knowledge.

It takes ten years to train an auditor, the same investment in time and complexity as training a pilot.

Bolting in the new Audi workshop is done with electric wrenches. Tightening data is uploaded and saved for 15 years.

The Audi A6 (C5) officially rolled off the assembly line at FAW-Volkswagen in Changchun on 6 September 1999. The vehicle was lengthened by about 100 mm to 4,886 mm, and the wheelbase was 90 mm longer than the global version. It was a great market success for over five years with more than 200,000 sold, until it was replaced by the new C6 in April 2005. Financial returns more than covered the additional investment to lengthen it.

The lengthened version of the Audi A6 (C5) became the benchmark for this class of cars. BMW developed the 5 Series L model for the Chinese market in October 2006. Sales in 2007 increased by 61 percent after it was lengthened by 140 mm. Mercedes-Benz also lengthened the new-generation E-class. But that's another story.

The Audi A6 won the approval of Zhongnanhai, especially when a bulletproof version was provided. This ensured the status of Audi as a high-end executive-class car, and it was branded as an official car.

3. Accord: Growing by leaps and bounds

One franc purchase: Peugeot goes home crestfallen

Guangzhou Peugeot, which was started in the same year as Shanghai Volkswagen, prospered for a short time. However, by 1996, when annual sales of the Shanghai Santana exceeded 100,000 vehicles, Guangzhou Peugeot had 10,000 vehicles it couldn't sell. The Peugeot 505 was a success in Europe. It had been modified since coming to China, but it seemed to have a curse on it. There was more product backlog than could be kept at the factory, and space rented at the local Huangshan airport was full.

Control of the joint venture was held by its Chinese partners. Guangzhou Automobile Company (GAC) held 46 percent; China International Trust and Investment Corporation (CITIC) held 20 percent; Peugeot of France held 22 percent; BNP Paribas 4 percent; and International Finance Corporation (IFC) held 8 percent. Both sides felt that GAC was weak in manufacturing and wondered if it could make and sell a European car while increasing localization. It was a big challenge even if they pulled together. What strange bedfellows they made, always worrying that the other side was trying to gain advantage. The situation just went from bad to worse.

Even today the Chinese side seldom acknowledges its own role in the failure of Guangzhou Peugeot. The Chinese partners had absolute control with 66 percent of the shares, but always placed their own interests first. The directors held meetings accusing managers who spoke French as weak and abandoning their principles, making it difficult for the management team to do anything.

Peugeot's top management lacked the strategic vision of the decision makers at Volkswagen, having no confidence in the Chinese market. Peugeot's 22 percent stake in Guangzhou Peugeot was not a cash investment, but a technology transfer and partial equipment investment, so it profited little from its interest. Peugeot shortsightedly put off proposals to increase capital and expand production, insisting on selling component parts to make money.

The Chinese side bore the burden of raising working capital and carrying the debt of the enterprise. Guangzhou Peugeot was constantly in deficit and relied upon the Guangzhou Municipal Government to order banks to give it loans. A local bank governor said that he would panic whenever the mayor invited him to morning tea. By 1997, the company was heavily in debt with accumulated losses of 2.96 billion yuan and was forced into bankruptcy. Its loans were all backed by the Guangzhou municipal government. When it went bankrupt, it took the municipal government's reputation with it.

The Standing Committee of the Guangzhou Municipal Party Committee met on 27 April 1996 to discuss the future of Guangzhou Peugeot, and painfully resolved to replace its partners and let Peugeot withdraw.

When Peugeot learned that Guangzhou was already negotiating the takeover with other auto companies, it was well aware of the pressure on Guangzhou City as a result of the bankruptcy meeting. Peugeot still stubbornly refused to withdraw from the venture, even though the city promised to cover the 500 million yuan overdraft for imported parts and a significant share of the transfer expenses.

Government authorities made it clear that all the loans would be guaranteed. But for a business that had been run so badly, it would only give Peugeot 1 franc for its equity. The Peugeot representative signed the agreement on 31 October 1997. CITIC, International Finance Corporation, and BNP Paribas also withdrew at the same time. Each received a 1 franc coin.

Honda gets its entry card for US$200 million

After weighing the cost of entry to the Three Majors and Three Minors and agreeing to some harsh conditions, Honda Chief Executive Munekuni Yoshihide put down US$200 million to take on the mess at Guangzhou Peugeot. With this, the Guangzhou Honda miracle began.

Replacing Peugeot was a shortcut into the Chinese automobile industry. Honda wasn't the only one waiting at the gates of GAC.

BMW, Mercedes-Benz, Fiat, Ford, and Mazda were all willing to take over, but they balked at the 3 billion yuan debt. GM Opel and Hyundai went furthest in their discussions.

At the very last minute, when a choice was about to be made, Honda unexpectedly showed up and submitted a proposal in January 1997.

Dongfeng Motor Group, formerly SAW, brought Honda to Guangzhou. The State wanted Dongfeng to reorganize the hapless GAC. Dongfeng and Honda discussed using the abandoned Huizhou Panda Car Project factory to manufacture components. GAC agreed to compare offers.

Honda soon came through with a complete proposal. Authorities felt that GM's project in Shanghai, which was still under construction, had an uncertain future. And that Hyundai, which was just emerging in the global auto industry, was far weaker than European and US companies. When Honda's submission was compared, Hyundai and Opel were eliminated.

Authorities in Guangzhou and Beijing agreed on the choice of Honda. Guangzhou Automobile Group, Dongfeng Group, and Honda Technology Co., Ltd. signed a joint agreement in Tokyo to implement the Guangzhou Automobile Project on 13 November 1997. Today's Guangzhou Dongfeng Honda Engine Company and GAC Honda Automobile Company are products of that agreement.

I interviewed Honda Chairman Munekuni Yoshihide at Honda's headquarters in Tokyo in 2004. As General Manager, he led the talks on the Guangzhou Honda project. Munekuni remembered all the faces of the opposing negotiating team. The one he couldn't possibly forget was Guangzhou Deputy Mayor Zhang Guangning.

The three biggest difficulties in restructuring Guangzhou Peugeot were the 2.96 billion yuan of debt, 1.11 billion yuan in fixed assets, and a 5,000 vehicle stockpile. Zhang Guangning astutely raised the cost of Honda's entry from 450 million yuan to 830 million yuan by selling the 1.11 billion yuan in fixed assets at a premium of 1.61 billion yuan to the new joint venture (Honda's 50 percent share was 800 million yuan). This meant that Honda paid a total of US$200 million (1.6 billion yuan). Additionally, the partners would each pay off

500 million yuan of the debt without compensation, and would make a one-off clearance payment of 500 million yuan for the stockpiled cars. GAC, which had been overwhelmed with debt, had suddenly cut its losses.

Munekuni was under tremendous pressure to sign the agreement. It must have been the same pressure that Carl Hahn of VW and Jack Smith of GM had come under. Giving in for the sake of compromise allowed Honda to defeat its opponents and enter the Chinese market, a first for a Japanese company. He won a big prize for Honda.

Progressive market-oriented development

Guangzhou Honda Automobile Co., Ltd. was established on 1 July 1998 in the old factory at Huangpu left by Guangzhou Peugeot. Kadowaki Koji was the first General Manager. He was nearly 60, but still had the perseverance of an old thoroughbred.

Kadowaki hung a chart in the office showing progress on 353 projects, including stamping, welding, painting, fitting out the four major assembly workshops, and public infrastructure. Facilities were completely renovated or torn down and rebuilt. New equipment was installed to produce the high-quality Honda Accord. With the additional 500 million yuan investment, the 353 projects were completed on time and a 30,000 vehicle production line was ready in nine months.

The first 2.3-liter Accord rolled off the production line on 26 March 1999. Honda's Technical Director Yoshino Hiroyuki attended the ceremony and said that renewing Guangzhou Honda and producing a superior sedan were big challenges. It was also a first for Honda.

In the 6+3 structure of the global industry, Honda was an outsider, without the financial strength and assuredness of the established companies. Its partner, GAC, had already suffered a setback in a joint venture and couldn't afford to lose again.

Guangzhou Honda had pledged to raise local content to 40 percent within 18 months. The schedule was tight for Lu Zhifeng, the first Executive Deputy General Manager. A key to achieving the target was Dongfeng Honda Engine Co., which was built at the same time.

A group of Japanese parts factories, optimistic about the strength of China's auto market, also followed the automaker to Guangdong.

When the first Accord came off the assembly line eight months later, audits by the Ministry of Machinery Industry and the General Administration of Customs confirmed that the Accord had reached a localization rate of 45 percent. Guangzhou Honda was ranked first for quality among Honda's 17 overseas factories.

The Accord sedan was in great demand when it came on the market. The 10,000 vehicles produced that year didn't come close to meeting demand. Zeng Qinghong succeeded Lu Zhifeng as Executive Deputy General Manager at the end of the year. He and Kadowaki basked in the glory of Guangzhou Honda's market-oriented, low investment, and high output development. Annual output leapfrogged from 10,000 to 200,000, from the first cars in 1999 to Kadowaki's retirement in 2004.

The general rule in China's auto industry was that a company must invest at least 6 billion yuan to be able to break through 100,000 vehicles. Some large manufacturers had invested more than 10 billion yuan.

The dirty factory left by Guangzhou Peugeot was transformed into a garden-like factory by Guangzhou Honda.

When Guangzhou Honda reached 120,000 in 2003, it had only invested an additional 2 billion yuan. When production capacity reached 240,000 in 2004, additional investment was only just over 5 billion yuan, a third of the normal figure. All of this money came from Guangzhou Honda's profits. Shareholders hadn't invested another penny.

Zeng Qinghong said communications and trust were the basis for Guangzhou Honda's success. Unanimous decisions on market demand, localization, costs, and services were achieved through communications. All decisions had to have the signatures of both Chinese and Japanese directors. From investment, procurement, and personnel to operating expenses, both had to countersign before anything could be done.

4. The New Century crosses the Pacific

The wisdom of SAIC personnel

How did the car industry in Shanghai develop after annual output reached 100,000 vehicles? The State agreed to Shanghai's engaging in another mid- to high-end car project as early as July 1994, but several problems had to be resolved, including capital and localization rate.

SAIC began discussions to choose between GM and Ford in 1995.

Hu Maoyuan, who had already become SAIC Group Vice President, said, "Negotiations are spontaneous when you have a down-selection and it is a good way for a student to choose an instructor. We set as the condition a mid- to high-end sedan that could be retrofitted as an MPV[5] (multi-purpose vehicle), and which we could keep on developing." Both GM and Ford brought out new products in the second half of 1995. For GM it was Buick's New Century, and for Ford it was its US best-seller, the Taurus. GM's asking price was lower, and it agreed to build a joint venture R&D base in addition to the factory. SAIC chose GM.

---

[5] MPV is short for multi-purpose vehicle is a type of car generally favored by families due to a more practical interior than a regular hatchback, often coming in five- and seven-seat forms.

SAIC and General Motors signed a letter of intent for technology transfer and a joint venture at GM's headquarters building in Detroit on 31 October 1995. The technology transfer fee for three GM models, including Buick's new Century, was US$48 million. The total investment for the factory, including the Pan Asia R&D Center, was US$1.52 billion. SAIC organized a negotiating team, called the Pudong Car Project Team, with Hu Maoyuan as team leader. GM sent Phil Murtaugh who had a background in international operations.

Hu Maoyuan recalled, "I first met Phil Murtaugh at the 1996 Detroit Auto Show. GM Vice President Rudy Schlais introduced us. Murtaugh was friendly and I felt I could trust him. Once I took three leaders, Zeng Peiyan, Lu Fuyuan, and Jiang Yiren, on a visit to the US. We flew on a chartered plane. Murtaugh received us and helped us with our luggage himself. We stayed on good terms for more than ten years."

"The most difficult part of the negotiations," Hu Maoyuan said, "was the product liability clause."

Chinese regulations required that a product liability clause be included in the technology transfer agreement. The US had strict product liability. There are two kinds of compensation, one for direct loss which requires direct compensation, and one when a defective product is knowingly produced and can be traced back to the parent company. The GM negotiator said that for this US$1.5 billion project, they couldn't accept unlimited risk.

To break this deadlock, Hu Maoyuan visited the Economic and Trade Commission's Articles and Law Division. They said, "This is a worldwide problem and it can't be resolved. The only thing you can do is to give up this clause." But the feedback from the Commission's Technical Department was that this article had to be adhered to because it protected Chinese interests.

Hu Maoyuan lay awake all night thinking about it. They must fight for it no matter what because it protected SAIC's interests as a technology purchaser. He calmly analyzed GM's concerns. They were willing to pay if they were responsible for problems, but they were afraid of putting their entire company at risk.

When he thought about it that way, it suddenly became clear. He proposed limiting compensation to US$48 million, which was within the amount of the technology purchase price. GM's concerns were eliminated and the two sides agreed. Hu Maoyuan was praised for creatively solving the problem.

GM's determination to win

Jack Smith, who had served as President and CEO of General Motors since 1994, was of medium height, rather chubby and had an easygoing, attentive, albeit somewhat aloof manner.

GM was on the verge of collapse with annual losses of US$15 billion when he took charge, but it regained its position atop the automobile world through drastic cost-cutting and structural reforms.

With sustained high growth in the Chinese economy, the outlook for the world's last big car market gradually became clear. When news

The low-key, level-minded, and visionary GM Chairman Jack Smith receiving Chinese reporters in his Detroit office after he signed the contract with Shanghai GM in January 1997.

came in the mid-1990s that China was establishing another joint venture in Shanghai, Smith was determined that GM had to win it.

GM went all out. Ms. Yang Xuelan, well-known in the US public relations field, joined GM as a Vice President. Ms. Yang was the foster daughter of Gu Weijun, a diplomat in the early years of the Republic of China. She was elegant and sophisticated. She organized a full-scale public relations campaign that went beyond cars. Later, in her residence at Shanghai's Old Jinjiang Apartments with first-rate contemporary Chinese paintings hanging all around, she recalled for me that Sino-US relations had fallen to a low point in the early 1990s when the US was selling F-16 fighter aircraft to Taiwan and had invited Lee Teng-hui to the US. It wasn't a good time for a US company to go into China. She pressed on nevertheless, mobilizing GM's Chinese employees to use all possible channels to communicate with friends and relatives in government, science and technology, culture, and universities. The Committee of 100, composed of Chinese-American elites, was also actively engaged. In addition, influence was exerted on local congressmen, through GM's US dealer network, to improve cooperation and eliminate misunderstandings between the two governments.

The hard work paid off and GM won the competition, overcoming its many powerful competitors.

I covered the North American International Auto Show (NAIAS) shortly after the joint venture agreement between SAIC and GM was initialed in January 1997. On the day we were to leave Detroit, we were told that Smith would meet with Chinese reporters in his office.

GM China Public Relations Director Ms. Hao Cuixia, who accompanied us, said excitedly, "I've been with GM for 25 years and I still haven't been into the chairman's office." This unusual courtesy showed how much China was on GM's mind.

GM's headquarters was still in a 1927 granite building. The majestic pillars at the front door and the beautiful paintings on the dome of the lobby gave the building a sense of history. The 15th floor was the most important part of the building. That was where GM's top leadership had their offices, and it could only be reached by a dedicated elevator.

Smith's office was at the end of the corridor and you could see him working at his desk through the open door. I was told his schedule was arranged in five minute increments. After a few minutes, Smith stood up and walked to the door to shake hands with us.

The meeting was relaxed. Smith wore a white shirt and a dark tie. Leaning against his desk with arms crossed, he asked for our impressions of the past few days.

Smith had achieved amazing results at GM, and people compared him to Alfred Sloan[6]. But he wasn't complacent, saying he had only completed half of his mission.

The meeting went ten minutes longer than scheduled. Smith gave each Chinese reporter a ballpoint pen with the GM logo and a group photo was taken. "Things are going to get better in the new year," he said. He made an OK sign with his fingers. It seemed like a wish for GM and for us.

Smith announced the establishment of GM China in Beijing. Being the first major auto company to set up a subsidiary in China was an innovative move. Smith appointed a deputy, GM's Vice President Rudy Schlais, to take charge of the Chinese business and serve as President of the Chinese company.

Shortly after assuming the position, Schlais announced five principles for its China company: a long-term commitment to cooperate with China's auto industry; providing advanced technology; training local Chinese management and technical personnel to international standards; participating in all aspects of China's automobile production, including complete vehicles and parts; and, helping China build an internationally competitive automobile industry. These commitments were received favorably by the Chinese. I had heard senior Chinese officials cite them more than once.

---

[6] Alfred P. Sloan Jr. (23 May 1875–17 February 1966) was an American business executive in the automotive industry. He was a long-time president, chairman and CEO of General Motors Corporation who helped GM grow from the 1920s through the 1950s.

Everything is based on the interests of joint ventures

SAIC Chairman Lu Ji'an and GM Chairman Jack Smith signed the Shanghai GM joint venture agreement in Beijing on 25 March 1997.

Hu Maoyuan, Phil Murtaugh, and Chen Hong were the principal and deputy general managers. Shanghai General Motors Corporation held a groundbreaking ceremony in Jinqiao in Pudong on 25 June that same year.

GM Chairman Smith and SAIC Chairman Chen Xianglin drove the first Buick New Century off the assembly line on 17 December 1998. People were in tears as the new car broke through a paper screen. Building a modern car factory in just 21 months hadn't been easy.

The schedule was so tight that it was only possible to construct the framework of the buildings. Awnings were put up when it rained, and people kept on working as they trampled through the mud. It was just like the Daqing Oilfield work spirit that we used to talk about.

There were conflicts within Shanghai GM in its early days, because employees came from different countries and had different cultural backgrounds.

Hu Maoyuan went to a board meeting in Detroit in the second half of 1997. Kept awake by jet lag, he thought about how to create an enterprise culture at Shanghai GM. He proposed four points in his report to the board:

1. Have unified goals and be in step with one another.
2. Cooperate for mutual success, not unilateral success.
3. Agree on some simple rules. "I can't be at loggerheads with Murtaugh and we can't express opposition to each other in front of subordinates. We can't pass our conflicts upwards. We have to close the door when we quarrel and observe two minutes of silence afterwards. And we can't say anything when we are angry."
4. Use the 4S concept for cooperation — joint venture partners should learn from each other; set interests as a priority; standardize actions; and be flexible and pragmatic.

Chinese and foreign employees taking a photo with the first Buick New Century that came off the assembly line on 17 December 1998.

Putting a joint venture's interests first is a yardstick. When problems occur, fairness comes from putting the joint venture first and not saying which side benefits. Looking at the long-term contributes to sustaining joint venture development, and is crucial to it.

Many years later, Hu Maoyuan and I talked about our own past ideological struggles. We had all said that the interests of the party, the interests of the State, and the interests of the people were priorities. Wasn't it rightist to say that interests of joint ventures were a priority during the anti-rightist era? Then, I realized that the profits of joint ventures could be allocated proportionally in accordance with the investment. It was in everyone's interest for the joint venture to be profitable.

Everyone valued the mutually profitable corporate culture. Murtaugh had a good relationship with the Chinese workers. When we talked, he often told me what the workers wanted. I joked that he was like a union boss. He had the people-oriented philosophy of US companies. Why did Shanghai GM propose value management as a way of meeting consumer needs? It was because they absorbed good things from others.

GM had an open attitude at the outset of the joint venture. It formed a team of consultants to impart the experience of its global sales experts to its Chinese partners. In this enlightened way, it introduced a market competition mechanism, which influenced the whole Chinese automobile industry.

I contributed to a three-part GM TV series in 1999 at the invitation of GM Global Vice President Ms. Yang Xuelan. When my script was accepted, I was impressed by the fact that the US producers just added pictures and music, without changing a word. The series was broadcast as an GM advertisement on CCTV and important local stations. In independent surveys two months later, GM's popularity among the Chinese public had increased significantly. On MTV's China on Wheels, which I produced, there was even a choir accompanied by a symphony orchestra. It was broadcast on CCTV's Weekly Songs during the 50th anniversary of New China. It sounded magnificent and heralded the arrival of China's automobile age.

Smith's vision

I flew to Brescia in Italy in June 2000 as the only Chinese media representative at GM's seminar on the future of GM and the global automotive industry in the 21st century. Senior GM executives and 50 of the world's most influential automotive media people and financial analysts were there.

Brescia sits beside Italy's largest freshwater lake. It wasn't hot although the sunshine was bright. The water and sky were an expanse of blue and there was an occasional breeze. The red tiles and yellow walls of the castles were hidden among green foliage. Perhaps the globalization of the auto industry had been too fast. Choosing such a quiet, peaceful place let everyone reflect calmly and think about the future.

The seats in the venue were arranged elliptically, on three levels. It was a bit like the arrangement in the Roman Senate. There were no

In the Shanghai GM's vehicle body workshop, the chassis and the body are welded together, known as "marrying" in the industry.

assigned seats and no designated seats for leaders. There were blank cards on the table at each seat. Participants randomly chose their own seat, and filled in their names on the cards.

Everyone sat down. GM CEO Rick Wagoner, the moderator, asked each person to introduce himself. GM Chairman Smith sat unobtrusively next to a US journalist in the back row.

Wagoner made the keynote speech. He said, "We have ushered in the biggest wave of mergers since the 1920s. It's a major change. Mergers in the 1920s took place within industrialized nations, where hundreds of automobile factories merged into several large companies within each country. The restructuring is global this time. The goals today are to achieve growth by adapting to new market segments, to increase economies of scale, and to take control by forward-looking actions."

In the question time that followed, I was surprised to hear one Wall Street financial analyst after another expressing serious doubts about GM's recent investment in China — demand for mid- to high-end cars was only a few thousand and shareholders' money was being put at risk. Smith and his team were under tremendous pressure.

"I can't seem to avoid this subject," Smith said calmly. "There are 1.3 billion people in China. Not everyone can afford a car, but some have reached a fairly high consumer level. The population of China's coastal areas is about 400 million, and the purchasing power of the region is similar to that of most Central European countries. Poland has a population of 40 million, and total car sales were 500,000 last year. In China, which is much larger and much stronger, sales were only 650,000. This is worth thinking about. Returns will be considerable for the first company which enters a regulated market with low entry costs. However, one of our principles is to bring the strongest technology and management to China and do the best we can there."

Smith said to the skeptical financial analysts and the media, "China is commercializing fast, and Shanghai has changed remarkably. Roads and infrastructure are being built 24 hours a day. As the first US auto company to go into China, many of our standards are now becoming new standards there."

No one could have imagined what the Shanghai GM project meant for the future of General Motors.

I had my last interview with Smith, who was about to retire at the end of a 41-year career at GM, when he came to Shanghai in the summer of 2004. At his farewell party with the GM Shanghai employees, China Chairman Murtaugh toasted him saying, "Many of you have contributed to GM in China. However, only one person deserves to be called the founder of our business here, that person is Mr. Smith."

03 Working in someone else's garden: An impression of Kadowaki Koji

Swearing off alcohol

Kadowaki Koji was a wine drinker.

Kadowaki, 62 years old, was about to retire from the position of General Manager of Guangzhou Honda, where the Chinese staff affectionately called him the Old Man. On the way back from the Geneva Motor Show in April 2004, I flew to Guangzhou to see him before he left. That night, we drank a small flask of Huadiao wine[7] together.

A year before, the new Honda Accord sedan, launched worldwide, had come off the Guangzhou assembly line. Production jumped from 10,000 vehicles to 59,000, and people thought he could have a drink to celebrate his success. However, he refused, saying "Forgive me, but I've made a promise to myself. We're building up to production of 240,000 cars per year. I won't drink any wine until we get there."

Having reported on the automobile industry for many years, I knew that an annual output of 240,000 cars was incredible for a Chinese factory producing mid- to high-end cars. Later, Kadowaki told me that when Japanese people made up their minds to do something, they had

[7] High quality yellow rice wine from Shaoxing, Zhejiang Province.

a custom of giving up something they liked, so as to remind themselves to reach their goal.

Kadowaki brought about one final coup. Following the Spring Festival in 2004, Guangzhou Honda changed over to its new 240,000 vehicle production line. It became the largest car company in China with the largest single production line. The same year he stopped drinking, the company created three more miracles: its output doubled, supply couldn't meet demand, and its economic results were the best in the industry.

Realizing his Chinese dream

Kadowaki had studied Chinese at university. He joined Honda when he graduated, and worked there for 39 years. He worked in factories in the US, Belgium, and Canada, and at Honda headquarters in Japan, becoming a specialist in market development. He feared fate wouldn't take him to China, but in his fifties, he was appointed Honda's chief representative in China. We met for several interviews in his office in the Xingfu Building in Beijing.

He was low-key, reserved, and didn't say much, but he listened to what other people said. His eyes watched you fixedly behind his round glasses. He was involved in Honda's final negotiations with Guangzhou Automobile Company. When Guangzhou Honda was established in July 1998, Kadowaki became its first General Manager. In his own words, he finally had the luck to realize his Chinese dream.

When he took the job, Kadowaki and an older colleague stood at the factory gate looking at the dilapidated site left behind by Guangzhou Pugeot. He said to his colleague, "You and I better work hard and do a good job, because it looks like this is the last place we'll ever work."

Kadowaki knew some Chinese sayings. "Practice what you preach" was his motto. He began with cleaning the whole factory. He announced at a meeting of all the personnel, "I'll be responsible for imports and exports (by which he meant the canteen and the toilets), and I'll leave everything in between to you." Two weeks later the factory looked like new. A staff canteen, where a thousand people could eat, went into operation on the third floor. The toilets were unblocked, and a mountain of garbage on the factory site was cleared away. Today, Guangzhou Honda is a world-class factory in a garden-like setting.

In January 2003, Guangzhou Honda launched the new upgraded Accord, which was higher in quality, but 40,000–50,000 yuan cheaper. I wrote an article praising them for starting a wave of price cutting in the auto market. Before Kadowaki retired, Guangzhou Honda produced the small, sleek Honda Fit. Kadowaki said, "Our aim in launching the Fit was so that young people buying their first car could have a world-class one. We set a target price of 100,000 yuan, and worked backwards from there. That would normally be hard to do, with purchase of parts and production costs, as well as taxes on imports. But people can do a lot when they have a goal and work together. Despite the pressures, we delivered on the price promise, and without loss."

Communications are essential

When I asked him if he had planned it all beforehand, he shook his head. "It was a time of rapid growth in the market, and we responded quickly, that's all." Laughing at himself, he said, "When I went back to Japan two years ago with a plan to start producing 240,000 vehicles a year, many of my colleagues at headquarters couldn't believe it. They said in alarm, 'Is Kadowaki dreaming?' But common sense prevailed and they agreed."

Kadowaki smiled as his Chinese dream had been satisfied. He told me that when he started as General Manager, he hadn't thought that Guangzhou Honda could be what it had become. The reality was more beautiful than he had dreamed. I asked him, "With your wealth of experience working overseas, what is the secret of success in multinational cooperation?"

He thought for a moment and said, "Effective communications are essential. As an overseas Japanese, I always pay attention to one thing — I'm working in someone else's garden, so I have to respect other peoples' culture and customs and make the garden richer and more beautiful."

Although globalization is making the world flatter, understanding, respecting, and adapting to each other's cultures is a prerequisite for success, even if it is just a soft one.

Once on a program on CCTV, the host asked for advice for Chinese carmakers working overseas. I passed on what Kadowaki had said. "Remember the Japanese man's words. Chinese auto companies also have to figure out how to work in other people's gardens."

Chapter 5

New Sedans Eagerly Knock at the Door

oreigners might not understand that in the second half of the
20th century, government departments and even the highest
policy making levels had the final say on which companies could
produce cars. Getting permission was impossible for ordinary local
state-owned enterprises, let alone private ones. At the turn of the century, a few companies appeared and were infatuated with making cars.
They were seen as car mad and like catfish stirring up the stagnant pond
of the Chinese auto industry. No one could foresee that years later, this
new generation would become the backbone of the Chinese car industry and the forerunners of indigenous brands.

Night was falling, and the autumn wind was cool. Li Shufu and
some media friends were walking in the streets near the Asian Games
Village in Beijing. "The Geely is ready to go. All that's left is the right
to produce it." Looking up at the waning moon, Li Shufu sighed dis-
piritedly. In July 2001, just as China was about to join the WTO, the
Geely hadn't been included on the State Economic and Trade
Commission's product list.

Half a century of high tariff barriers had given the government
enough reason to show concern for the weaknesses of state-run automobile companies. Gaining a six-year grace period for the Chinese auto
industry was a hard-won achievement of the ten-year WTO entry
negotiations.

However, the scope for protection was limited. According to the
strategic plan for automobile development issued by the State
Economic and Trade Commission, the key protection targets were the

three state-owned enterprises — FAW, Dongfeng, and SAIC. Taking advantage of the grace period, the government protected these first. Industry outsiders, especially private ones which were just emerging, were strictly excluded. They closed the front door to the tigers, but let the wolves in by the back. That's how greatly they feared competition.

With indifference and lack of compromise, the new generation who were making cars under great difficulty were always in danger of facing delay in getting approval for their products.

I couldn't sit by and watch. I had to do something.

1. The new generation hangs by a thread

WTO: Calm and terror

The Central Economic Work Conference was held in the Huairentang building in Zhongnanhai on 15 November 1999. As usual, only two or three media organizations like Xinhua attended. As Director of the Xinhua News Agency Economics Reporting Office, it was my job to write the press release for the conference.

At the same time, Chinese and US government representatives were negotiating China's entry to the WTO. A tired Premier Zhu Rongji rushed from the WTO negotiations to attend the opening ceremony. He said simply that the negotiations had reached a critical point, and were going on day and night.

He left hurriedly as General Secretary Jiang Zemin was making his keynote report.

Jiang Zemin analyzed the situation facing the Chinese economy in the coming century and talked about the three characteristics of globalization. Firstly, rapid development of science and technology provides the impetus for economic globalization. Secondly, multinational corporations drive economic globalization. Thirdly, industrial adjustment no longer happens within a country, but on a global scale.

We had been talking about globalization for more than ten years, but people didn't really understand what it was. That was probably the most authoritative statement I had heard by any Chinese policymaker.

Zhu Rongji reappeared on the rostrum at 4 pm, and announced that an agreement had finally been reached at the WTO negotiations. China's entry was no longer in doubt. In his subsequent speech, he revealed that Shi Guangsheng and Charlene Barshefsky[1] were the delegation heads. His report of the behind-the-scenes negotiations made the senior Chinese economists hold their breath at times, and laugh aloud at others.

An informal reception was held in Huairentang that evening. To everyone's surprise, Jiang Zemin, Zhu Rongji, and other members of the Standing Committee attended. Uncharacteristically, Zhu Rongji took to the stage and sang some Beijing Opera. He was relaxed and happy.

Economic reform had reached a stalemate and it was hard to break through the complexities of old systems and interests. Frankly speaking, the help of a "Shaman" was needed to promote China's WTO entry. The rules of globalization and marketization were used to break down the old system, introducing competition mechanisms and market rules, and taking the Chinese economy into the profitable world of globalization.

In the words of Long Yongtu, China's chief representative at the negotiations, "Old habits are hard to break, and we couldn't do it on our own." It took a vigorous and effective external force to push us to do what we wanted to do.

The wolf was at the door. But not everyone could decipher the calm reached by the strategists.

State-owned enterprises had undergone 12 years of reforms, but their losses were still growing. I went to the National State-Owned Enterprise Reform Work Meeting every year, and I knew all the representatives from the local committees. There was an atmosphere of helplessness. They joked that the war of resistance against Japan had only taken eight years, but this could take longer.

---

[1] Charlene Barshefsky (born 11 August 1950) served as United States Trade Representative, the country's top trade negotiator, from 1997 to 2001. In 1999, she was the primary negotiator with China's Zhu Rongji, laying out the terms for China's eventual entry into the World Trade Organization in December 2001.

A closed mindset made a lot of people believe that as markets opened and multinational capital flowed in, the faltering state-owned economic system would not be able to bear it, and noncompetitive industries would be driven out of the market. The automobile industry was terrified by that prospect. With tariff protection of 180 to 220 percent, and cars costing three or four times as much as abroad, it didn't know anything about competition.

The annual output of China's auto industry was 300 billion yuan in 2000. That drove annual output of related industries to 200 billion yuan, employing 10 million people. The government couldn't give up on it. Protection of the auto industry was one of the difficulties that made the WTO negotiations drag on so long. The concern was that once the world's automobile products came in, the Chinese automobile industry would be wiped out. This was so ingrained that Chinese representatives spent a lot of effort to get a six year grace period during which tariff barriers would be gradually reduced.

In the final agreement, tariffs on imported cars would gradually be lowered until 1 July 2006 when the rate would be 25 percent for cars, and 10 percent for parts and components.

Keeping the catfish alive

As we were about to join the WTO, I began communicating with the new generation outside the industry and sharing my ideas with them. I was impressed how determined they were to make cars for ordinary people. They were the best and most savvy group of Chinese entrepreneurs. They were like catfish stirring up the murky pool of the auto industry with completely new market mechanisms. They were also the fortunate survivors of a life-or-death assault against a rock-hard system. With their high IQs and experience, Zhan Xialai, Yin Tongyao, Yang Rong, and Li Shufu knew how to do the impossible. They were car enthusiasts, and they were the backbone of future indigenous brands.

The Ministry of Machinery Industry was dissolved in the late 1990s. The management functions of the automobile industry were largely carried out by two departments. The first was project approval through the State Planning Commission to decide whether a car company could legally exist and the second was management of the automobile product catalog by the State Economic and Trade Commission to decide whether a car model could be sold on the market. The former was like issuing a household registration for a family; the latter was like a birth permit allowing the family to have a baby.

I referred to management of the automobile industry as a living fossil of the planned economy, from which the market and competition were excluded. State Council Order No. 24 in 1997 clearly stated that no more car projects would be approved, regardless of their capital structure.

China's registered automakers had grown to 120 in a few decades, more than the total number in the rest of the world. Eighty percent of them produced less than 10,000 vehicles a year, and more than a dozen between zero and ten a year. Because they were on the catalog, they lived on comfortably like spoiled Manchu nobles at the end of the Qing Dynasty. It seems unthinkable now.

Outside of that system, the privately held Jili (Geely), Qirui (Chery), Huachen (Brilliance Auto), and Yueda had obtained their production permits through gray channels by buying shell companies and converting shares, using item code 6 for buses, and refitting vehicles other than sedans to enter the field of sedan production. However, properly obtaining the item code 7 permit for sedan production was far beyond their reach.[2]

As China was about to join the WTO in 2001, the State Economic and Trade Commission issued an Automobile Strategic Development Plan, supporting the three major automobile groups with capital investment, project approvals, mergers, and restructuring.

The Automobile Management Department had the dual responsibility of managing the industry and revitalizing state-owned enterprises.

---

[2] In 1988, China issued GB9417-88 "Automotive Product Modeling Rules," using simple Chinese Pinyin letters and Arabic numerals to indicate a domestic company's enterprise code, vehicle type code, main characteristic code, product serial number, enterprise customization code name, and so on. In this standard, the vehicle types were classified for the first time. "6" stood for passenger cars (including light buses) and "7" stood for cars, and specifically sedans. GB9417-88 has since been abolished, but this classification management of automobile products was used for a long time.

They still seemed to be doing things in the same old way. Joining the WTO was just around the corner, but they still wanted to exclude private auto companies from a typically competitive industry, to protect the three big state-owned enterprises. The State changed from catalog to bulletin management of auto products, and the State Economic and Trade Commission issued the first Bulletin of Vehicle Manufacturers and Products on 22 May 2001. But the quasi-legal car companies were not on it, despite their high quality and low price, their environmental and safety standards, and their popularity. They couldn't produce or sell new products, and were at an impasse.

Automobile industry veteran Chen Zutao commented that China's auto industry hadn't grown in 50 years. It had monopolized the industry, keeping others out.

I was indignant at this indifference by the authorities. I wrote in an article entitled "What Are the Crimes of the Quasi-Legal Sedans?" in my column "Car Talk for Laymen":

> People have publicly said sacrifice is inevitable. Let's go back and review our lessons on competition. We're out of time, and time and our assets can't afford any more losses. Quasi-legal car makers, with their own legally earned money and under great hardship, are making the sedans that Chinese people have been dreaming of. What are their crimes? Why are they being sacrificed? Quasi-legal sedans makers are refusing to give in, saying: "US companies can make cars in China. I'm a Chinese company, why can't I make them? When you make cars with state money, if you lose you can just exchange debt for equity and carry on. Why can't people who make them with their own money do that?"
>
> The State bails out large state-owned enterprises to get out of trouble and makes national teams. But they must be broad-minded enough to let outsiders compete. At least keep some as sparring partners. If you don't have anything to do with them, they will still survive. It's wishful thinking that you will survive global competition when the grace period is over.

I passed this article to an old friend, State Planning Commission Director Zeng Peiyan. He was a strong promoter of Chinese cars, and especially family cars. He forwarded my article to the Planning Commission and expressed support for quasi-legal cars.

2. The Chery: A surrogate birth

A daredevil's persistence

Anhui Chery was one of four companies that were being shut out. It was an unconventional company with an unforeseeable future.

I flew to Nanjing on the evening of 7 November 2000, then jolted along the road to Wuhu for two hours. The scholarly Zhan Xia, Secretary and Mayor, was waiting for me at the Tieshan Hotel, though it was late. As the regional party and government leader, he was under huge pressure to make cars.

I was impressed by the company, which is located on a vast field of grass in the Wuhu Economic and Technological Development Zone. The large grounds were reserved for the company's expansion, but were rented to a lawn turf company, which gave the factory a green space while adding to its income.

Qirui didn't have a general manager and it was Vice President Yin Tongyao who saw me. The first thing he said was, "We're like a band of desperadoes here."

Four shareholders, namely Anhui Investment Company, Anhui Insurance Company, Wuhu City Construction Investment Company, and Wuhu City Development Technology Investment Company all under Anhui's local financial governance, raised 1.772 billion yuan to establish Anhui Automobile Spare Parts Company, the predecessor of Chery Automobile, in 1997. Wang Yang, who was Wuhu Party Committee Secretary, gave the go-ahead. In March that year, Chery broke ground in the Wuhu Development Zone. The plant was finished a year later.

As a maker of spare parts, Chery naturally started with an engine. They bought second-hand Ford engine equipment from British DP for US$45 million. It was installed by a British company, which had no previous experience in China. For example, the power requirement of a European factory was 450 volts at 50 Hz, and for China it was 380 volts at 60 Hz. The British were at their wits end about it. But they were arrogant and saw China as backward. Yin Tongyao had been the director of the FAW Volkswagen Jetta assembly workshop, and successfully relocated the Volkswagen Westmoreland factory equipment. They got rid of the undisciplined British in Novemberworked around the clock for 150 days, and made their first engine on 18 April 1999. After that, they designed and customized four major workshops.

The first Chery came off the assembly line on 18 December 1999 without its birth permit. The entrepreneurial team took a photo with the new car. First left in the back row is Yin Tongyao. Second right in the front row is Assistant General Manager Jin Yibo. The foreigner at the back, named Altmann, worked with Yin at FAW Volkswagen. He went to Chery after he retired and worked there for 11 years.

Yin Tongyao recalled: "I'm from Chaohu and was assigned to FAW when I graduated from Anhui Institute of Technology. I later became director of the FAW-Volkswagen Jetta assembly shop. When Chery was starting, Secretary Zhan visited FAW and FAW Director Geng introduced me to him. They were both from Anhui. Zhan invited me to dinner that evening. I didn't want to go at first and suggested we don't dine but just talk. I was impressed by his enthusiasm for cars. He asked me how the factory should be built and I told him my ideas and drew diagrams. I talked on and on, the prospect making me unable to stop.

In the end, I made up my mind to join Chery and bring a few like-minded colleagues with me. Most of them were from Anhui. They were called the Eight Warriors. I became Deputy General Manager and Director Geng gave me a lot of support. He sent veteran experts like Lin Ganwei to teach me management, operations, and technology."

Yin Tongyao was about forty, with a youthful face and a full head of hair. He was elected as one of the top ten outstanding young people in FAW-Volkswagen. He was plain-spoken and very determined. There was no general manager at the time, Yin Tongyao said, and they were waiting for someone to fill the post.

Chery developed its first sedan at the Fuzhen Company in Taiwan on 14 February 1998. They sent people to observe each stage, from design to production of dies and molds. A screw body[3] and stamped parts for 60 vehicles were made in Taiwan and shipped to Wuhu in October 1999. The first Chery came off the assembly line on 18 December 1999.

A birth permit in exchange for shares

I went to the Wuhu Communist Party Committee, a simple 1970s building, for an afternoon of conversation with Zhan Xialai. Although he had studied Chinese language and literature, you could tell he was a car enthusiast like Yin Tongyao. He told me cars had given him nothing but trouble. He wasn't the kind of cadre who would just use the company as a springboard for his career. I could identify with that.

As the real spirit behind the scenes at Chery, Zhan Xialai, who was promoted to chairman by the four shareholders, guided the company from a macro perspective. He summed up its spirit as "Think ahead, work hard, run fast, and say little." He told me that Chinese people could make the best and cheapest family car because it has lower labor

---

[3] Under a functional build approach, manufacturers typically assemble body components into prototype car bodies or "screw bodies." The logical progression is to build up small sub-assemblies (builds) as separate screw bodies, and then to assemble the individual builds into the whole body.

and development costs than the international industry and a market of more than one billion people. After 20 years of reform, a complete car component network has been established, fully adapted to Chinese conditions. He said Chinese car companies were well-structured, without the high welfare, strong trade unions, and labor-capital conflicts that existed in developed countries.

Tiggo, an early SUV developed by Chery and equipped with Chery's own turbocharged engine, has been its star product for many years. The latest generation Tiggo 8 is shown here.

However, Chery still couldn't get onto the national car catalog. That's why they asked me to Wuhu.

The Chery sedan was born out of the A-class car that VW's subsidiary brand SEAT was producing. I thought it was very well-made. It had a 1.6-liter engine and cost 80,000 yuan. It was 50,000 yuan cheaper than the Jetta. I drove it from Wuhu to Nanjing in the rain. On the road to the airport, I got it up to 150 kilometers per hour, and it handled well.

Chery had an annual production capacity of 60,000 vehicles. In accordance with local Anhui policies, the cars could be produced and sold within Anhui, and all the taxis on the streets of Wuhu were Cherys. In addition, Chengdu, the pioneer city of family cars, allowed the good quality low-priced car to be sold there with Anhui license plates. But the Ministry of Public Security put a stop to that, and the factory filled up with unsold cars.

Zhan Xialai told me one way to obtain legal status was to go into a joint venture with Shanghai, transfer 20 percent of the shares, and become a junior partner in the SAIC Group in exchange for getting on the catalog. However, he didn't want Chery to be a steppingstone for technology transfer and have its planning done by others. He wanted me to write an internal report to the central leadership about what Chery was doing to get into the Chinese car industry, and to convince the authorities to give it official status.

I had some understanding of the Politburo Central Committee's thinking about macroeconomic development because I went on study tours and took part in meetings with them. I didn't think there was much chance, but I wanted to give it a try. I attended the Central Economic Work Conference at Zhongnanhai on 8 November 2000. Premier Zhu Rongji talked about the 10 key tasks of the Chinese economy in 2001. His thinking was that as China entered the WTO, the most important thing would be to revitalize key state enterprises, and approval of new construction projects would be strictly reviewed, suspended, and delayed.

As soon as I left Huairen Hall that evening, I called Zhan Xialai to tell him what I thought about this new direction. I suggested that Chery should join the SAIC Group immediately, and not pin its hopes on my article moving the leadership. I said it was a life or death choice for Chery. It would be worth it to become a new Shanghai brand. Zhan said Shanghai had been waiting for their decision which, with my call, they could now make. With authorization from the shareholders, he called SAIC President Hu Maoyuan to inform him that Chery had decided to join SAIC.

Chery got approval to join SAIC at the beginning 2001, and the company changed its name to SAIC Chery Automobile Company. SAIC was very accommodating, giving Chery room to develop and declaring them in the catalog. Chery was allowed to sell cars in the Chinese market, and was the first quasi-legal car maker with the right to exist.

The Chery lived up to expectations, with its graceful appearance, reliability, and realistic pricing. Sales exploded the following year, increasing 50 percent almost every month, until 40,000 had been sold, making it the most popular Chinese car on the market.

3. Geely gathers strength
Always looking for new opportunities

My book *The Lure of Family Owned Cars in China* became a bestseller at the beginning of 1998. When he read it, Li Shufu sent someone to ask my advice about Geely's plans to build a car. I said it wouldn't be easy. Cars are a big-investment, high-output industry, especially in China. The risks were enormous, and the large investment could be lost.

Li Shufu mentioned this in public many times. But he was not one to play by the rules and inclined to do what he thought was right. The streets were full of the Geely Haoqing (HQ) and Merrie (MR) in two or three years and the quality wasn't as bad as I had thought. The Geely passed the same collision test as a Mercedes Benz at Tsinghua University. It taught me a lesson, showing it was possible to build an affordable 30,000 yuan car.

Li Shufu was born in a wealthy village in Huangyan in Zhejiang Province in 1963. People there knew how to make money. When he graduated from high school, he ran a photo studio, recovered precious metals from industrial waste, and ran a refrigerator factory. He was making a lot of refrigerators and freezers by the end of the eighties, and a long line of delivery trucks at the factory gate waited to take them away.

Li Shufu was able to avoid the risks that could have capsized him. Like many private entrepreneurs in Zhejiang, he was a loyal fan of CCTV's News Network program. After the turmoil of the Tiananmen Incident in 1989, people debated whether private enterprises were socialist or capitalist. Li Shufu and several partners gave the township land, plant, and equipment for a factory, then went off to Shenzhen to study at the university. He saw Agriculture Minister He Kangzhong encouraging and supporting township and village enterprises on News Network at the end of 1991. He was perceptive enough to realize that national policy had altered and a new round of economic development was about to begin. He packed his bags and left.

"Whatever I did, someone would immediately copy me and undercut me," he recalled. "So I thought, it would be harder to copy if I made cars. We were making aluminum-plastic panels, the brand was Geely, and the market was very good. We had some money in hand, so in 1993, I decided to try something bigger and make cars."

Li Shufu went to the Director of the Municipal Economic Commission and asked for approval. The Director smiled and said it was impossible. Then he went to the Provincial Machinery Department in Hangzhou. They also said it was impossible because they had no direct connection to Beijing. They couldn't build a car factory without State approval, so the path was closed to them.

Li went back to thinking about motorcycles. He knew he wouldn't get approval for that either, so he found a motorcycle factory that was on the verge of closing down, bought somebody else's brand, and started to make motorcycles, but they didn't sell well.

Then something happened that turned things around. The company sent an employee on a purchasing mission, and he borrowed a Taiwan-made Koyo scooter. There was an accident, for which the company paid compensation of 20,000 yuan and kept the damaged scooter. No one in the country was making scooters then. Li Shufu found that it had more plastic cover panels than a motorcycle. Motorcycle production was already strong, the plastics industry was developing, and there were imported Yamaha engines. These came together, and after feeling their way forwards for more than a year, Geely started developing its scooter.

He said, "The market was very hot when the scooter came out. Three or four hundred thousand vehicles a year, and we had the biggest market share in the country. The number of manufacturers increased, and the price dropped from 20,000 yuan to 2,000 yuan. We also acquired the Linhai Lucheng Motorcycle Factory and started using the Geely brand."

Despite this success, Li Shufu never forgot his intention of making a car. He used the appellation of motorcycles to buy 800 mu of land in the Linhai Economic and Technological Development Zone. He decided to go all out and build a car factory.

Coming down to earth from Mercedes-Benz and Boeing

Li Shufu wasn't a rigid thinker. The first car he built in 1997 was a high-class copy. The Mercedes-Benz E series had just been superseded. Li Shufu bought several Mercedes-Benz E280s, disassembled them, and made drawings of them.

He bought an FAW Red Flag sedan chassis with the front and rear axles, engine, and gearbox. He made a fiberglass body, copying the shape of the E280, and bolted it onto a Red Flag chassis. He made a trip to Hong Kong to buy a Mercedes-Benz steering wheel and front and rear lights.

That's how his first car was built. He proudly showed off his "Mercedes" around the city.

"A lot of people wanted to buy my car," he recalled. "If the authorities hadn't disagreed, I could have gone into large-scale production right away. The problem was that in the eyes of many people, I seemed to have broken the law."

"My machinery department director told me I couldn't do it, but I didn't give up. I went to the Machinery Industry Ministry in Beijing about making motorcycles, which I already knew how to do. When I showed them a photo of my 'Mercedes,' they said something like, 'What will happen to state-owned enterprises if you produce this car?'"

"Of course, I couldn't argue with them, but I didn't give up. I asked them what car I could produce? I asked them to show me a way forward."

"They didn't give me any directions, just wouldn't let me do it. But I figured out one thing, to make a car it had to be in the national catalog. There was no hope of getting in the car catalog, but there were all sorts of companies in the bus catalog."

"The catalog was a book. I flipped through it and found that one of the Deyang Automobile Plants which produced buses had stopped production, but it was still in the catalog. So, through a friend in Sichuan, I asked them if I could buy their place on the catalog and they said it could be done cooperatively. We paid tens of millions of yuan and I called it Sichuan Boeing Automobile Company."

"Now we were in the car industry, but what model would we produce? The people in Beijing told me I couldn't compete with the big companies, and it had to be a small compact car. According to their ideas, we made one based on the Xiali. The body was a nice shape made of fiberglass."

The Haoqing 6360, legally skirting the line as a bus, came off assembly at the "Boeing" branch in Linhai in Zhejiang on 8 August 1998. It cost 60,000 yuan, which was competitive with 100,000 yuan domestic economy cars at the time.

Li Shufu prepared a 100-table banquet and invited officials at all levels. But a private enterprise making cars was too radical, and most people didn't dare go. Only Zhejiang Vice Governor Ye Rongbao came to offer his congratulations. Seeing ninety empty tables laden with food gave Li Shufu another taste of the hypocrisy.

When it was reported in the media, another problem came up. The US Boeing Aircraft Company went to the State Economic and Trade Commission. Although there were no cars among its products, Li Shufu wasn't allowed to use the Boeing name. Later, Li Shufu bought out Deyang's shares, transferred the catalog status to Linhai, and produced the Haoqing at the Linhai factory.

Then he bought the shell of Ningbo Auto Tractor Factory and acquired 1000 mu of land in Beilun in Ningbo. He started building on 18 August 1999. The company has since changed its name to Zhejiang Geely Automobile. Geely's other environmentally friendly, electronic fuel injected four-cylinder sedan the Ningbo Merrie was launched in 2000. At a price of only RMB 65,800 yuan, it set a record as the lowest price for cars of its type in China.

Two Geely models made Li Shufu famous: the Ningbo Merrie and the Geely Haoqing.

Longing for transparency

Li Shufu was a fearless industry pioneer who always startled everyone, but his predictions often came true.

Just before China joined the WTO, Li Shufu said in an interview with *China Entrepreneur*, "I long for transparency."

In 2001 he went to a high-level industry forum held at the China World Hotel in Beijing, of which I was one of the promoters. Speaking of the changing global car landscape, he said major companies like GM, Ford, and Toyota would all fail. When people slammed Geely's original model as a copy, he said, "There's no mystery about cars. Cars have been two sofas on four wheels for a hundred years. No one has made a five-wheeled car, so they're all copies."

Because of his grassroots origins and outspokenness, he was excluded from the mainstream industry. As WTO entry came closer, the State Economic and Trade Commission issued its Automobile Strategic Development Plan, proposing development of cars focused on the three major state-owned enterprises, and not approving any new projects.

Although the car was selling, it had an illegal category 6 designation as a bus. Geely and other quasi-legal companies tried to get into the catalog and sought equal treatment in market competition. Geely even started a Division Two football team to increase its visibility.

Li Shufu went from office to office within the province and in Beijing looking for a way forward. He didn't care about the coldness and sarcasm he encountered. He pleaded, "Just give us a chance, and even if we fail, at least you will have shown fairness."

That year, car products changed from catalog to bulletin approval. Li Shufu put a lot of hope on the July Economic and Trade Commission Bulletin. He thought the two new Geely cars would get the go-ahead. But yet again it came to nothing. So, when others told him Geely wasn't on the list, Li Shufu didn't even have the courage to open the newspaper where the Bulletin was published.

To help him, I prepared a cover story "Life or Death for Li Shufu?" for the October 2001 issue of *China Entrepreneur*.

I commented that the government's policy of avoiding competition would eventually kill Li Shufu, and it would also kill the Chinese automobile industry. It would only take five or six years of protection to lose it all!

In the interview for the magazine, Li Shufu said, "People say I'm like Don Quixote, but I won't end the way he did. I hope China joins the WTO soon. Geely and the Big Three are as different as heaven and earth, but those differences cannot be allowed when we join the WTO. Geely isn't afraid of competition, even from multinational companies. There is room for us because no multinational can produce low-end 30,000 yuan cars. There are a lot of benefits in joining the WTO.

Others can use our capital, talent, technology, and components, and we can use theirs."

"What I'm longing for is transparency", he said. "It's a central issue for Geely. If we can obtain production rights, then capital and technology are not a problem. Mid- and low-end cars will be mass-produced by Chinese companies in the future, supplying Chinese and foreign markets, that's for sure."

The State Economic and Trade Commission suddenly issued new car bulletin on 9 November 2001, and the Geely JL6360, with an engine of its own design and stamped metal body, was finally on the list. A private company, determined to make good cars that ordinary people could afford, was finally given the go-ahead just one day before China joined the WTO and 26 years after the first car joint ventures were approved.

China's indigenous brands welcomed WTO entry with price reductions. Tianjin Xiali, which had been working hard with Geely, lowered its 100,000 yuan price, announcing an astonishing new price of 30,000 yuan. Geely responded with a lower price of 29,800 yuan. "China Joins the WTO" and "Li Shufu Gets In" appeared together in the same newspapers.

I wrote a column congratulating Li Shufu. I said getting a birth permit hadn't been easy, but a child had been born, and we hoped it would grow up healthy. Growing to be big and strong would be more difficult, but that's what happened.

4. Brilliance: The elephant in the room

The Zhonghua alternative

I first saw the Zhonghua at Tsinghua University in early winter 1999, when I met Brilliance Chairman Yang Rong.

Graceful and sleek, the black model was in no way inferior to the world's most successful designs. Its stylish curves, enclosed headlights, and narrow side windows gave it a dynamic, forward-looking design. A badge with the character 中 (Zhong) in a ring was at the front. I still remember what I felt when I saw it. I was stunned.

CHAPTER 5 | 163

A lot of prototypes were being developed, and most of them were small and quite crude. The Zhonghua, a mid-level car which was the work of Giugiaro[4], was a harbinger of Chinese car design.

Yang Rong, a neat man of medium height with big eyes and bushy eyebrows, looked bright and expressive.

"Mr. Li, we invited you here today because the development of the Zhonghua was inspired by your article." He told me with sincerity and modesty.

In 1993, Brilliance had just taken over production of the Jinbei van and a lot had to be done. Yang Rong, an active thinker, saw a report about modern Korean cars I published in *Outlook* magazine, which gave him the idea of making cars. He gave copies of my article to the directors at a board meeting. A national brand, high level research and development, making use of resources and personnel in domestic and foreign markets, and having their own engine; all the methods used by Hyundai and summed up in the article could be applied on Brilliance. The board was persuaded to make the important decision to go into cars. This began the only legend in the history of Chinese cars directed by financial operations, which integrated global resources, a high starting point, and indigenous R&D in a domestic automobile.

Brilliance started engineering the Zhonghua sedan in June 1997. Yang Rong summed it up by saying, "I have China in my heart, but I'll use the world's resources."

Yang Rong hired Italian designer Giugiaro to take charge of the US$62 million design. Giugiaro knew just what was needed. Chinese people liked the feeling of riding in a sedan. They revered simplicity and comfort, and the length of the body was the lucky number 4888.

---

[4] Italdesign Giugiaro S.p.A is a design and engineering company and brand based in Moncalieri, Italy, that traces its roots to the 1968 foundation of Studi Italiani Realizzazione Prototipi S.p.A by Giorgetto Giugiaro and Aldo Mantovani. Best known for its automobile design work, Italdesign also offers product design, project management, styling, packaging, engineering, modeling, prototyping, and testing services to manufacturers worldwide.

Brilliance was the sole investor, and therefore owned all intellectual property of the Zhonghua. Yang Rong positioned the Zhonghua as medium- to high-end, with parts and components sourced globally, at half to one-third the price of an imported car.

The Zhonghua stood out among indigenous brands. The KUKA welding line, Dürr coating line, and Schenck assembly line in its Shenyang factory were the same as in the new GM and Volkswagen plants in Shanghai. When BMW insisted on a joint venture with Brilliance, it was firstly because of their market mechanism, but also because of their world-class production line.

The Zhonghua came off the assembly line in Shenyang on 16 December 2000. Yang Rong said to me, "Mr. Li, will you be my consultant from now on?"

Listing in New York

Brilliance kept a low profile for nine of its first ten years. Although sales of the Jinbei and the Toyota Sea Lion made up 60 percent of the market for similar vehicles, Brilliance's profits were second only to Shanghai Volkswagen among domestic automakers. People became aware of Brilliance with the advent of the Zhonghua. One financial magazine called Brilliance "a mystery." Yang Rong bore it all with a smile.

Yang Rong is a native of Anhui, with a PhD in Finance from Southwestern University of Finance and Economics. He made his first fortune in the early Chinese stock market, and opened the Huabo Finance Company in Hong Kong.

He had a chance meeting with Zhao Xiyou, Chairman of Shenyang Jinbei, in 1990. Zhao was the first in the auto industry to try changing the shareholding system, and had sold Jinbei shares to Zhongnanhai. However, a few years later, just under half of the total 100 million shares had not been sold. Yang Rong bought the remaining 46 million shares for one yuan each. It was registered stock, packed in dozens of cartons, and airlifted back to Shanghai, where it was placed in the basement of Building Seven of the Municipal Party Committee's Donghu Hotel, with an armed police guard.

卯 珺

二〇〇一年十月十九日

| FIRST BOSTON | October 19, 1992 | CITIBANK, NA
55 WALL STREET
NEW YORK, NY 10015 | No. 063112 |

| New Issue | | *74,400,000.00* | 1-8/210 |

PAY **Seventy Four Million Four Hundred Thousand and 00/100 .... DOLLARS**

TO THE
ORDER
OF   Brilliance China Automotive Holdings Limited

Aan : / Cc

AUTHORIZED OFFICIAL

⑈063112⑈ ⑈021000 810 53506⑈

US$72 million was raised when Brilliance was listed in New York. Yang Rong sent a copy of the bank check to the author.

The 46 million shares, if listed at a tenfold premium, were 460 million yuan. But they were only wastepaper if they didn't get onto the market.

Shenyang Jinbei brought in technology to produce the Toyota Sea Lion van, but lacked start-up funds. Yang Rong was optimistic about the growth of this new product, and took a loan of US$12 million. Brilliance, in its capacity as a Hong Kong business, established a joint venture with Jinbei — the Brilliance Jinbei Bus Company — to produce the Toyota van.

Yang Rong showed his financial expertise by listing Shenyang Jinbei on the Shanghai Stock Exchange in July 1992. It became the biggest overseas stock on the exchange. He converted his 46 million Shenyang Jinbei shares, and paid off his outstanding loans.

However, Yang Rong's outstanding coup was listing Brilliance China Automotive Holdings Co., Ltd. in New York in 1992. Perhaps that was too avant garde. Many officials and economists doubted him.

Yang Rong had extraordinarily keen judgment. Huangfu Ping's articles were published in January 1992 while Deng Xiaoping spent the holiday in Shanghai, and Yang Rong thought it might trigger a new wave of economic reform. He told Zhao Xiyou he wanted to list overseas.

Yang Rong always picked the hard thing to do, and then did it. He chose the most strictly regulated overseas capital market, the New York Stock Exchange. He took several young assistants to New York and went to investment banks on Wall Street, like First Boston and Merrill Lynch, to discuss the feasibility of listing. That shook up Wall Street. But the spark that started the project was Deng Xiaoping's Southern Tour speech. The 14th National Party Congress was in the fall of the year. The wave of reform promoted by Deng Xiaoping would be handed over to a new generation of leaders. Yang Rong wrote a research report to the highest levels of the Central Government, showing it was feasible for Chinese companies to be listed on the New York Stock Exchange, and that China would still follow the path of economic reform post Deng Xiaoping.

Later, Brilliance acquired a 51 percent stake in Jinbei Bus through an exchange of stock with Shenyang Jinbei.

Yang Rong registered a company in Bermuda with a stake held in Jinbei Bus: the Brilliance China Automotive Holdings Limited (listed as CBA on NYSE). Brilliance held 100 percent of the shares.

It was impossible for a Chinese private enterprise to be listed in the United States. The China Financial Education Development Foundation was established as a corporate legal entity by Brilliance, the People's Bank of China Education Department, China Finance Institute, and Huayin Trust in May 1992, to be an equity holder for overseas listed companies. The registered capital was 2.1 million yuan, of which 2 million was funded by Brilliance and 100,000 by the Education Department of People's Bank. Before the listing, Brilliance transferred the controlling interest of CBA to the China Financial Education Development Foundation as a "donation."

That way, all the CBA documents and materials were standardized and transparent. They were reviewed by US Certified Public Accountant Anton & Chia and the listing was approved on 18 September.

On 9 October 1992, CBA was successfully listed on the New York Stock Exchange, New China's first overseas listed company. CBA was also the first stock listed in New York by a socialist country. The controlling shareholder was China Financial Education Development Foundation, holding 55.7 percent. CBA issued 5 million shares of common stock, at an offer price of US$16 per share. It rose by 25 percent on the first day of listing, becoming the most active stock on the New York Stock Exchange.

A ceremony for the transfer of funds obtained from the listing was held in Beijing on 28 November. The successful listing made US$80 million. Yang Rong and representatives of the three big investment banks which had assisted with the CBA listing were met by a member of the Central Leadership at Zhongnanhai. After a group photo, Yang Rong stayed behind and handed over a sample stock certificate to the leader. When he inquired who the stock belonged to, Yang Rong's full-blooded answer was "the nation."

After that, the State-owned Assets Supervision and Administration Commission (SASAC) wrote to the China Financial Education Development Foundation saying all CBA equity and stock returns were owned by SASAC. An investment of US$15.3 million was made by the State Administration of Taxation and invested in CBA on behalf of SASAC. The Foundation bore no liability and did not own the benefits (meaning it was in name only). SASAC also gave the Foundation US$3 million to establish financial education awards.

The investment from SASAC mentioned in the letter was the donation from Brilliance. In other words, the Foundation was the nominal holder of the Brilliance donation. Brilliance donated the equity of CBA to the State. As the recipient of this asset, SASAC reissued its documents, entrusting Yang Rong to manage CBA. Yang Rong was young, enthusiastic, and patriotic.

The 14th National Party Congress was held three days later. CBA exemplified the persistence of China's third-generation leaders in Reform and Openness, and was a model for many Chinese companies which would list overseas later.

Wet hands dipped in flour

Money was the scarcest resource in the Chinese auto industry for a long time. Yang Rong was first to connect the financial pipeline to the auto industry.

Brilliance never got any government funding or tax breaks. However, Yang Rong always seemed to have as much money as he needed.

Brilliance General Manager Su Qiang, who had single-handedly brought Jinbei Bus from loss to revival, said, "The money all comes from common practices in markets, from gathering the original capital, from capital markets and from business profits.

Zhao Xiyou said, "People say Brilliance just pretends to make cars so as to make money on the stock market, and it will eventually take its money and clear out." In fact, Brilliance's investment, as well as the funds obtained by overseas listings, were invested 100 percent in Jinbei Bus and did solid business. The two major shareholders agreed at the outset not to divide profits, but keep reinvesting them. Not a penny was taken out.

Brilliance borrowed from banks, but its deposits and loans were balanced and its credit was excellent. When the chairman of a commercial bank inspected the finances of Jinbei Bus, he was surprised and said that if all companies were like Brilliance, China's finance industry would be at no risk.

I wasn't sure if it was coincidence or inevitability, but at the turn of the century, Brilliance got capital from its cooperation with five major auto companies. Sanjiang Renault, Jinbei GM, Toyota Sea Lion, and Mitsubishi Engines were all in joint ventures or partners with Brilliance. The joint venture with BMW was the biggest surprise.

The tallest tree in the forest would be blown down first. Someone in a report to the Central Work Committee for Large Enterprises said Yang Rong was a financial speculator who was embezzling state-owned assets. Yang Rong told me: "Speculating in the financial market is the easiest way to make money. Capital turns into factories, equipment, and car designs. Money just piles up there, and like flour sticks to your wet hands, there's no way you can shake it off. You just have to keep toiling away until the cars come out."

You can only win if you know how to change

"You can only win if you know how to change, but if you keep to your usual way you will lose." Yang Rong and I were talking about his motto in the 28-story Brilliance Building near the Huangpu River Bund. He said that this truth from Sun Zi's Art of War had been tried and tested in stock market battles for ten years.

Regarding the intellectual property rights of the Zhonghua, the joke at the time was that Brilliance had property rights but not intellectual ones. An automotive industry leader once asked, "When is the next generation Zhonghua going to be developed?" Yang Rong calmly replied, "We already have the second and third generations, and the next generation will be competing with international cars of the same class."

However, the Zhonghua sedan was still waiting for approval. Yang Rong said that since the day he decided to invest in the Zhonghua in 1997, he never saw approval as an obstacle. He foresaw China joining the WTO, and when that happened, the issue of approval would be solved.

Yang Rong knew changes were coming in the automobile industry and that competition would come sooner or later. He was also a risk-taker.

While negotiations between Brilliance and BMW were proceeding, Yang Rong's thoughts jumped to another project that might be ten times bigger. He took a risk to cure the industry's "heart disease" by setting up an engine plant with an annual output of 1.5 million units, producing the world's most advanced turbocharged direct injection engines.

Insiders said it was a crazy dream. Whose technology would he bring in? Where would the billions of dollars of investment come from? To whom would he sell that many engines?

Work started on the Brilliance New Engine Plant in the Beilun Development Zone in Ningbo, Zhejiang in 2001, with an investment of 5 billion yuan. Compared with the restraints in Shenyang, Yang Rong liked Ningbo's freewheeling atmosphere. Brilliance signed a contract with FEV[5] from Aachen, Germany, for the engine and production equipment. The standard of machining was nearly the same as that of Mercedes-Benz and BMW. Pilot production began at the end of 2004. 500,000 units were planned for the first phase and 1 million for the second. The 1.8TFI turbocharged direct injection engine went into production in 2005. All products had the Zhonghua logo, with 60 percent going to Zhonghua, and 40 percent to British Rover.

The British Rover (MGROVER) was another of Yang Rong's big projects. The Rover factory built in Birmingham in 1877 was acquired by BMW in 1993. Due to operating losses, BMW withdrew and sold it to British Phoenix for £10. It was seen as the last wholly-owned piece of the British auto industry.

Yang Rong said, "BMW introduced Rover to me. They were doing badly. They had the technology to develop new products, but not the money. We hit it off from the start. The British Government and the Royal Family all supported it. It's tiring to be the only one chosen to carry the banner of indigenous development. With overcapacity of more than a fifth in the world's car production, my new idea was to recycle the world's assets, do mergers, and reduce logistics." He had that plan in 2001!

They negotiated to establish a joint venture product development company. Brilliance invested US$450 million, and Rover US$300 million plus technology. All funds were from overseas bank loans, with 50:50 equity. It was co-developed with Rover's four new 45 Series models and the new 25 Series, which were put into production in 2004 and 2005 respectively. A new 75 series came out in 2007. The contracts were signed on 12 March 2001. The Chinese and British would each

---

[5] FEV Germany is an internationally recognized leader in design and development of advanced gasoline, diesel, and hybrid powertrains and vehicle systems.

When the Zhonghua came out, people said that Yang Rong was a financial predator who speculated on cars.

produce under their own brand, with banks taking a £100 commission for each 25-series produced and £400 for each 45-series until their US$750 million loan was paid off.

At the same time, a 300,000-capacity automobile plant was set up in Ningbo, and all Rover models would be produced in China. 150,000 25 series and 100,000 45 series vehicles were produced in 2003. Rover stopped production of the 25 Series in the UK the next year, and 80,000 of Ningbo's vehicles were sold back overseas. It was sold as the Zhonghua domestically and the Rover abroad.

If things had proceeded smoothly, the Chinese car industry would be doing things differently today. We would have been eight to ten years ahead in the acquisition of overseas capital, and the introduction of the world's top technology.

Yang Rong runs away

I got the news at the beginning of March 2002. After an investigation at the highest level of the Central Government, Brilliance was being devolved to Liaoning Province.

Yang Rong would have been satisfied with a place to transfer some of his assets. However, CBA along with the entangled equity relationships of the entire Brilliance Group were transferred to Liaoning.

When this windfall first came to Liaoning, Yang Rong was treated courteously. But it was time to clarify property rights. Although Brilliance didn't have a penny of state investment according to Yang Rong, its overseas listing was a national resource and Yang Rong had committed to give the CBA shares to the nation. A 60/40 or a 70/30 split of the property rights was explored.

However, a storm arose because Brilliance was paying for the Ningbo engine project in installments and the Governor of Liaoning saw that as tunneling funds out of Jinbei. Things changed suddenly when Yang Rong was forced to flee to the US in May, and the whole of the Brilliance Group fell to Liaoning as a state-owned asset.

So, Brilliance became a state-owned enterprise in Liaoning Province. The Zhonghua sedan project and the joint venture with BMW were finally approved. The Zhonghua finally got onto the Bulletin in August 2002. But there was neither ability nor interest in Liaoning in looking after the Ningbo engine and Rover projects. When Brilliance changed ownership, the joint venture with Rover was abandoned.

Chapter 6

WTO Entry Creates a Market Boom

I greeted the new century four hours earlier than my relatives in Beijing on 1 January 2001. I was watching the final stage of the Asia-Pacific Le Mans Endurance Race in Adelaide, Australia. The downtown area was crowded with people, and as midnight drew near a loud countdown began. Bells rang in the New Year, and fireworks shot up and exploded. Everyone was cheering, clapping, and hugging, and strangers were wishing each other a happy new year.

China was at a new starting point. It had finally broken through its 50-year self-imposed barriers and joined the WTO, promising to abide by global trade rules so that it could enjoy unobstructed benefits. Another 50-year taboo was broken with the proposal of the Party Central Committee and the National People's Congress to encourage private car ownership for the first time.

Almost nobody thought joining the WTO and allowing private car ownership would become the major forces behind a 10 year boom in the Chinese auto market. China's auto industry ushered in its best and fastest-growing decade as it went from a bicycle kingdom to become the world's leader in automobile sales. The old fears were quickly forgotten.

1. WTO entry and family car ownership: Two great boosts

2001: China's Year of the Family Car

On my way to the North American Auto Show in Detroit, I flew home from Sydney on 2 January 2001. I was delighted to meet my old friend

Lü Fuyuan n the flight. As Deputy Minister of Education, he had just led a delegation to New Zealand with Beijing Vice Mayor Zhang Mao to attend a flame lighting ceremony for the World University Games to be held in Beijing.

On the long journey, we mostly spoke about our issue of mutual concern — cars. As the industry was about to face the test of joining the WTO, Lü Fuyuan said, "Globalization is a multi-polarization of politics and the economy with the EU, Japan, ASEAN, and North American trade areas. In order to globalize, China must have its own backbone industries and enterprises. Successful auto joint ventures must be based on profitability in accordance with China's long-term interests. We have a huge market and are entitled to profits and technology. People care about the success of Chinese cars."

He said, "China's auto industry shouldn't be underestimated. In fact, a lot had been achieved in the past 50 years. Talent was brought together and the market was nurtured in the large-scale build-up of the industry. During the Eighth Five-Year Plan, investment was 58.8 billion yuan, 80 percent of which was invested in 13 key enterprises. There was concentration and no duplication among the 13, so conglomerates could be gradually formed through mergers. The US, Germany, Japan, and South Korea followed the same path. If we don't deal with the WTO now, the industry will continue to be controlled by the government, and its breakup will be unavoidable. We don't have much time left, and we can't wait until the enemy is at our gates. Some people entertain the illusion of cleaning up the auto industry and selling it to foreign groups, letting others take control. That is not desirable for a big country like China. Not long ago, a Volkswagen director candidly talked about buying out a Chinese joint venture in the *Financial Times*. It woke many people in the industry up. In its current position of weakness, the Chinese auto industry holding less than 50 percent of the shares can be overruled by multinational companies who possess the technology and the product development. The dismissal of Chrysler executives when it merged with Mercedes-Benz is a lesson."

The following year, Lü Fuyuan was transferred to the Ministry of Foreign Trade and Economic Cooperation. He served as the first minister of the newly formed Ministry of Commerce in 2003. He died of cancer in May 2004. The automaker, who was widely respected in the industry, did not foresee that ten years after joining the WTO, China's auto industry would take the crown of the global auto industry. And now Chinese people had the dignity of a car ownership culture.

2001 was the beginning of Chinese family cars.

The Fourth Session of the Ninth National People's Congress closed on 15 March 2001. The meeting adopted the Tenth Five-Year Plan for National Economic and Social Development. For the first time, the statement "family ownership of cars is encouraged" was included. While only this very short statement appeared in the 20,000 word outline, a truly major market had been found in New China. The people's right to a car culture was elevated from Approved to Encouraged in the 1994 Automobile Industry Production Policy.

Ford Motor Company, which was still trying to get into China, finally formed an alliance and signed a contract with Chang'an Group, the base for China's weapons industry. The contract was signed in Chongqing on 25 April, establishing the Chang'an Ford Motor Co., Ltd. Ford had been arduously seeking a partner in China for 20 years, starting with SAW, leaping to Nanjing, then the battle over Shanghai with GM, before settling down in Chongqing.

The State Planning Commission announced the release of price controls on domestically produced cars on 10 May. However, the worsening price wars had shattered the old planning structure.

Shanghai GM launched the Sail on 8 June, six months after the market had heated up. It was a competitive platform from Opel South America with more than 100 local improvements. Sinking 100,000 yuan in a joint venture car was a first for Buick's second model in China. Aimed at families, it made an assault on the market with an Automatic Braking System (ABS) and other high-end options. Two years later, its sales exceeded 100,000 vehicles, capturing 27 percent of the small car market.

On 13 July, in Moscow, the International Olympic Committee voted by a large margin for Beijing to host the 2008 Olympic Games. It was a night of revelry in Beijing as traffic flowed spontaneously to Tiananmen Square.

On 11 September, a visa mistake left me in London for a week, waiting to fly to Prague to meet with other Chinese journalists. I returned to my hotel from the British Museum and watched the TV screen as the hijacked planes crashed into the World Trade Center in New York. I heard rumors that terrorists were threatening to attack London's financial district the next morning. But people didn't panic, they just helped each other, even me, a Chinese person with a problem. I felt a new era of change. China no longer confronted the world. Integration was an opportunity for China and for the world.

In November, BMW Chairman Dr. Joachim Milberg announced in Beijing that BMW Group had chosen Brilliance as a partner, and the State Council respected the choice. This move surprised my compatriots. If we hadn't joined the WTO, it would have been almost unthinkable for a non-mainstream automaker like Brilliance to distinguish itself among the other national players.

On 10 November, the Fourth Ministerial Conference of the World Trade Organization reviewed and approved China's accession to the WTO in Doha, Qatar.

That same day, I wrote a commentary entitled "Don't Waste the Grace Period," in my Xinhua News Agency column "Car Talk for Laymen." I wrote about the expectations of ordinary people, the government, business, and foreign companies after WTO entry, and the subtle differences between them:

> China's auto industry welcomes the WTO membership today. Its opportunities and challenges had been talked about for many years as though a wolf was at the door. Now, it has happened.
>
> One of the results of the negotiations is to give the Chinese auto industry a six-year grace period. It is a transitional period to strengthen an industry that is still in its infancy, though it has passed its fortieth year.
>
> People have no pity for the national automobile industry. They want the grace period to pass as soon as possible. Starting from now, they will buy nice, cheap imported cars, or they will wait for state produced car prices to be halved and fall to international levels. People don't care about a state car industry which has hit them with tariffs and high prices. Joining the WTO will force China's car prices to align with the international market, bringing Chinese people closer to a car culture.

The Government sees cars as a pillar industry that it mustn't give up lightly. Strengthening and expanding the industry within five or six years has become an unavoidable topic. An official of the State Planning Commission soberly said that now that we have joined the WTO, consumption of cars by ordinary people and participation in the industry by private enterprises and foreign manufacturers cannot be stopped. We won't get appropriate policies until we face up to reality. Changing the views and transforming the roles of government departments are the first steps to take in welcoming the WTO.

Auto companies in the front line of production and management feel a sense of urgency to seize the day. Chen Hong, General Manager of Shanghai General Motors Corporation, the largest joint venture in China's auto industry, said it's unlikely the price of domestically produced cars will change immediately. It's a process of improving competitiveness from engineering, procurement, logistics, and marketing to customer service.

Western countries welcome China joining the WTO because of the attractiveness of its market. We will have to give up part of that market, but not until capital and technology has been brought in. We have to use this to learn technology and management.

Of course, not all of the more than 100 current members of the Chinese auto industry will survive. The dynamic and market-oriented auto companies welcome the WTO, because being part of the Big Three will not be an inevitable condition for survival. The "spoiled offspring of Manchu princes," who have long since stopped producing cars and are living on industry protection by selling their catalog listings, are still doing well.

On 9 November, the day before China was joined the WTO, the State Economic and Trade Commission issued the sixth Vehicle Manufacturers and Products Announcement, and an unfamiliar model named Geely JL6360 was on the list. This was an unprecedented recognition of a private manufacturer. On the cusp of China's entry into the WTO, Li Shufu and Geely were accorded the same treatment as state-owned enterprises and joint ventures.

Private cars like the Geely, Xiali, Sail, Jetta and Fukang battled it out. 2001 was called the Year of the Family Car.

A white paper issued by Japan's Ministry of International Trade and Industry said China was the factory of the world. Chinese-made color TVs, washing machines, refrigerators, air conditioners, microwave ovens, and motorcycles ranked first in world market share. However, in 2001, China's motor vehicle production was 2.34 million, only 4 percent of the world market. There were only 700,000 sedans, an insignificant amount, but it became the basis for almost ten years of market growth.

The National Bureau of Statistics calculated that in 2001, China's per capita GDP was 7,543 yuan. At an exchange rate of US$1 for 8.27 yuan, the annual GDP per capita was US$912. Internationally, the car price/per capita GDP is called the R value, and an R value of 3 is the threshold for family cars. With an average domestic car price of 100,000 yuan ($12,000), only Shanghai, with an R value of 2.7, was within that threshold. Beijing, with an R value of 4.0, was on the borderline. The average R level in all provinces and cities nationwide was as high as 13.3. We were still far from private car ownership.

I wrote that the market was dynamic under the dual impetus of joining the WTO and family cars. Family car ownership in China was bound to begin in the large coastal cities. As GDP increased and car prices decreased, the R value would accelerate its drop below 3. Private cars would then move to the interior and grow rapidly.

China began to master everything from production to marketing after joining the WTO. When the Dongfeng Citroen Saina and Picasso came out, media and customers took part in the Dragon World Rally along the Silk Road from Beijing to Khunjerab Pass in South Xinjiang.

China's auto market characterized as a ten year "gusher"

China's car production in 2002, the year after joining the WTO, increased from 700,000 vehicles to 1.1 million vehicles, an increase of 53 percent! The global auto industry was stunned. The term "gusher" has been associated with the Chinese auto market for nearly ten years.

China, with a population of 1.2 billion, sold only 700,000 cars a year. There was no comparison to the US, which sold 15 million vehicles a year. There was a lot of room for growth. International automakers vied for this tasty morsel. Restrictions on multinationals were eased and almost all major automobile brands including GM, Ford, Volkswagen, Mercedes-Benz, BMW, Toyota, Honda, Nissan, Renault, PSA, Fiat, Hyundai, and Daewoo Kia, found joint venture partners in China.

People used to wonder who would come to China next, but there is no suspense now. All who should come have come. The contest has shifted to China, and there will be more intense competition for what's left.

New China's automobile industry is 50 years old. It has been an industry with trucks, not cars, and official vehicles, not private ones. However, globalization and family ownership caused unprecedented changes in 2002. It was a watershed and there were so many unexpected things, so much so that people couldn't keep up with them.

It was unexpected how quickly families would have cars. Ordinary people's long-suppressed purchasing power was released and markets exploded. During the Spring Festival, the street in front of the Beijing Asian Games Village Auto Trading Market was completely blocked by people who came to buy cars. Beijing's traffic department registered an average of 500 cars daily then, but there were 708 on 15 January alone. Companies began launching new products even faster. Over the course of the year, people were dazzled by more than 30 new models. The Sail, Xiali 2000, Palio, and Polo pioneered the compact family car sector. At mid-year, the Zhonghua, Bora 1.6, and Siena competed for the mid-range car market. And by the end of the year, the Vios, Sonata, Carnival, Golf, and Maxima went on the attack with advertising offensives. Car consumption was still far from mature, but now buying a car was like buying a color TV.

Then, a price war broke out. When Fiat's small Palio was launched in Nanjing on 29 January 2002, price cuts were announced on the Xiali 2000 and Sail, both of which fell below 100,000 yuan. Nanjing Fiat Chairman Flavio Ciappa and General Manager Mao Xiaoming met just before the Pallio's official launch to decide what to do. They announced the low price of 84,900 yuan for the basic 1.3-liter Palio that night. In the first 20 days of January, the Fu Kang New Freedom, Haima, Red Flag, Sail, and Antelope each lowered their price by more than 10,000 yuan. The Chongqing Alto and Geely set record low prices for new mainstream models of 38,000 yuan and 29,000 yuan respectively. People started opening their wallets. More than 80 percent of sales were for private cars. That was unthinkable two or three years ago.

Multinational companies were entering into joint ventures, instead of exporting vehicles to China. Almost all multinational automobile companies found joint venture partners here and became integrated. Mark Field, Executive Vice President of Ford's Premier Automotive Group[1] (a collection of high-end brands), told me that the per capita GDP of Beijing, Shanghai, Guangzhou, and Shenzhen was close to that of developed countries and their total population was the same as the UK. Meanwhile, the UK's car sales ranked second in Europe. Some big European and US companies had 10 or 20 year strategies for China and were looking towards potential markets around 2015.

The weak and scattered industrial structure that had plagued the industry for many years made breakthroughs in mergers and acquisitions in accordance with market rules. SAIC's 10 percent stake in South Korea's Daewoo created a precedent for the Chinese industry to enter

---

[1] The Premier Automotive Group (PAG) was an organizational division within the Ford Motor Company formed in 1999 to oversee the business operations of Ford's high-end automotive marques. The PAG was gradually dismantled from 2006 to 2010 with the divestiture of its constituent brands.

the global automobile capital market. Shanghai GM's restructuring as a 50 percent joint venture in Yantai Daewoo created a new model for domestic automobile mergers. Cooperation between Dongfeng Group and Nissan achieved new growth levels for sedans, and won new opportunities for its leading truck products. FAW merged with Tianjin Automobile, and then joined forces with Toyota. Car centers with unique strengths and characteristics, such as Shenyang, Guangzhou, Nanjing, Chongqing, Beijing, and Ningbo, became inseparable from the three major groups, forming a new 3+N pattern.

These explosive market conditions were only an introductory phase. It was only temporary. Private consumption of cars was far from mainstream in most small and medium-sized cities. In inland cities, ordinary citizens buying private cars could be seen as showing off their wealth, so they hesitated for fear that their work unit and neighbors might question the source of their wealth. I saw a program on Hunan Satellite TV where the host asked guests if their family car was really purchased by the family. The enthusiastic guests suddenly went silent.

People said real competition had not yet begun and companies were still jockeying for market position. I put forward a pyramid theory of the Chinese consumer market. The mainstream consumer base increases geometrically with every downward step in product prices. Popularization of family cars had to wait for sedans to continue down the pyramid in terms of models and prices. If it reached the township level at the base of the pyramid, the market would be huge. Prices of mainstream models ranged from 80,000 to 150,000 yuan in several big cities in 2002. Only at prices of 50,000–100,000 yuan would the impact be great enough to open markets further.

I compared family cars to fruits that can only be picked on tiptoe. But it's easier for people in cities where per capita GDP is US$3,000 than it is where it's US$1000, regardless of models and prices. I went to a conference in Tianjin. Most of the cars in the streets were little vans. I hadn't seen that for years. Although I was only a hundred kilometers from Beijing, I seemed to have gone back in time ten years. That showed me the inequality and disparity in China's growth.

China's Car of the Year

Local customs are a mirror of society and the economy. Beijingers used to ask each other, "Have you eaten?" when they met on the street. But in 2003, people asked, "Have you bought a car?" Cars, which used to be a dream, had become part of everyday life.

The Chinese car industry forgot its fears about WTO entry. The world had never seen anything like it.

The Chinese car industry had its highest growth rate in history in 2003. Total output was 2.03 million that year, an increase of 86 percent over the 1.09 million produced in 2002! This one year jump of one million was the first in 50 years.

The first Car of the Year was chosen in early 2003. Brilliance's domestic Zhonghua sedan received the honor. I was one of the judges of this groundbreaking selection which was organized by a media company working with the well-known German automotive magazine *Auto Motor und Sport* (AMS)[2].

In Europe and the US, a Car of the Year is awarded once a year by a third party other than manufacturers and consumers. The selection is generally made by professionals organized by the automotive media or the Automobile Journalists Association. They select one (and only one) car as the next year's most influential from among new domestic and imported cars. The important factors are technical innovation and value for money, so not many luxury cars are chosen.

Indigenous brands were still very weak in the Chinese market. In selecting the Zhonghua, the judges felt it was the first new model from an indigenous brand since the Red Flag and Shanghai in the 1950s. Its stylish design, advanced chassis, excellent handling, and affordability put it ahead of the others.

---

[2] Auto Motor and Sport, often stylized as auto motor und sport and abbreviated AMS or AMuS, is a German automobile magazine. It is published biweekly by Motor Presse Netzwerk's subsidiary Motor Presse Stuttgart, a specialist magazine publisher that is 59.9% owned by the publishing house Gruner + Jahr.

年度车2003

中华 *Zhong Hua*

中国人靠中国的力量和理念去大胆的设计这样一辆车。在中国历史上还没有。以16万元出台，它把价格真正压下来不留余地。作为公务车，它既能显示企业的身份，又能节省开支。如果中国的轿车都按这个价格定位并都有技术创新，那么中国汽车极具市场前景。从创新理念上讲它为现代中国汽车产业指出一条自主发展的道路。车身外形设计典雅大方，有冲击力。符合国际流行趋势，车身外观加工好。行走机构的设计出色，动力性能表现好。给人信心失分在于发动机匹配不好，噪声大、平顺性差。内饰加工粗糙。材料低档，与其中大型轿车的级别定位不匹配。这些都反应出一个新企业对质量，对产品把握的欠缺。很希一下做得完美。中华作为非常新的品牌，尤其是在价格上极具竞争力。但因为这个品牌不同于其它品牌有跨国公司作为基篱。在售后服务。产品质量可靠性等方面，用户都需要一个观察。认可的过程。

以赢得中国历来第一个"年度车"荣誉，其决定性的因素就是在最重要的"创新性"评价因素上表现突出。正如专家评委，新华社高级记者李安定先生所说。"它是一个理念上的创新。而理念创新比一个车型的创新更不得了。"

这个理念是什么呢？我们必须注意中华出现的背景：中国进入了市场经济的时代。加入了WTO大家庭。而华晨汽车是我国第一个在纽约和香港证券交易所上市的，国有资产控股的汽车企业。这辆车出

了我国传统国有汽车产业的氛围。它告诉我们。对于民族汽车工业的概念必须在现实的背景下，从新的角度理解。

中华轿车的创新性就是诠释了这种新的理解。市场经济，全球化之下的"民族性"应该怎样体现？它勾画了一条脉络。即资本控股－自有品牌－主导开发－全球资源整合－技术产权占有。其中的新意就在自有品牌跟整合全球化资源主导开发。

汽车是个产品，市场经济条件下，对

的崛起正是遵循了这个原则。中华车是继当年的红旗和上海之后。第一个我们有"控制权"的产品，所以说。中华是中国现代化汽车工业的一个里程碑。它指出一条"现代民族汽车工业"的发展道路。它的上市销售已经在中国汽车工业的史上写下了重要的篇章。

理念的创新无法独立存在。必须得到产品在实际水平上的支持。评委对中华的评价当然是以中华的技术、性能和质

The 2003 Car of the Year selection organized by *AMS Car Review* magazine was a first for China.

The selection of China's Car of the Year followed international practice. It was highly professional and impartial. On a cold day at the automobile test track of the Ministry of Transportation in the suburbs of Beijing, the judges tested each model on complex road conditions like dirt roads, corrugated and broken roads, and on sharp bends. Design, power, handling, safety, environmental friendliness, comfort, price, and driver satisfaction were scored, evaluated, and voted on.

The inaugural Car of the Year selection is something I'm still proud of. It was done better and followed the rules more closely than subsequent selections. The next year, a model with a new engine wanted to take part, but magazine Editor-in-Chief Xia Dong insisted that it wasn't a new model if only the engine was replaced. He refused to let it participate, doing all he could to maintain the integrity of the selection.

It was understandable that different media, with different audiences and value orientations, wanted to have their own Car of the Year. However, I wrote that I hoped one day, like in Europe, there would only be one Chinese Car of the Year.

There were more and more Cars of the Year when a TV station using its monopoly advantage took part. It selected multiple Cars of the Year to boost its advertising revenue. Cars of the Year increased to the point that a manufacturer might be able to attract attention by calling its product the non-Car of the Year.

The A League and the Four Flowers

The competition over A-class cars was like A-League soccer. It was the most important competition in the Chinese auto market. Dongfeng Citroen launched the Saina 2.0 sedan in Kunming in May 2003, and its domestic model, one of the so-called Old Three, finally had a replacement.

The name "Old Three" may seem a bit derogatory, but their historical successes were undeniable. They brought about the mass production of Chinese cars from scratch, and they established the mainstream status of domestic cars.

The Old Three didn't lose their popularity. 160,000 Santanas, 125,000 Jettas, and 85,000 Fukangs were produced in 2002. They led the more than 20 car brands nationally by a wide margin, and their sales increased by 30–40 percent in the first quarter of 2003.

The Old Three were introduced as standard A-class cars which were supposed to serve largely as official vehicles and taxis (private cars were still restricted then). They were each around 4 meters long with an engine displacement of 1.4 to 1.8 liters. As expectations of official passengers rose in the mid-1990s, the government approved the Audi A6, Buick New Century, and Honda Accord. All three were B-class cars, and senior officials got an upgrade. But their market size didn't compare with the Old Three which had reached economies of scale and had service outlets all over the country by the time the world's major manufacturers crowded into the market after WTO. Their status seemed unassailable.

The Santana was born in the 1980s. Although it wasn't a success for VW, it was popular in China for 30 years. I looked for it in the Volkswagen Auto Museum in Wolfsburg, Germany, but it wasn't there. There was only a similar station wagon labeled as a Passat B2, which was the great-grandfather of the Passat B5 that Shanghai Volkswagen later introduced.

The domestic Jetta was based on the second-generation Golf platform. It came out in the late 1980s and it was reborn after many changes in appearance and power train. I saw the Bora for the first time in Germany in 1999 and was told it was the fourth generation of the Jetta. The introduction of the Bora was very timely. It had several technical innovations and was known as a motorist's car. It had the largest increase in sales of A-class cars in 2002.

The Fukang came on the market at the same time as the Citroën ZX hatchback in 1992 and was the youngest of the Old Three. At first, many Chinese didn't accept the hatchback, saying it had a head and no tail. Shenlong (Dongfeng Peugeot Citroën Automobile Company, Ltd.) had to spend a lot of money on marketing to overcome this perception, benefiting the many popular hatchback models that followed. Shenlong's introduction of models reached a peak around 2003 with the Elysee, Saina, Picasso, and Peugeot 307. The Elysee was the successful result of domestic modification.

The market segments the Old Three had cultivated, together with their popularity and reliability, were valuable assets. Because of the diversity in the market, the Old Three remained popular because of their durability, low prices, and product updates.

In 2003, the biggest spoiler for A-class cars was Shanghai GM's Buick Excelle. The prototype came from Daewoo, which had just been acquired by GM. It was the first new GM Korea car model transferred to Shanghai GM and there were obvious cost advantages. However, the interior and chassis were redeveloped and remodeled by the Shanghai Pan Asia R&D Center. It was modern and elegant, and prompted GM Vice Chairman Bob Lutz to predict that the Chinese would surpass the Japanese in interior design one day.

Lutz flew to Shanghai to test drive the products of GM and its rivals in China at the Pan-Asian R&D Center, and to finalize the design of Shanghai GM's new car. Colleagues asked him what a car for Chinese people should be. Lutz replied, "The market segment is important, but cars for Chinese people have to be designed in China, not in North America. I've seen new, high standard models at Pan Asia. We have to listen to what Pan Asia says about cars designed for the Chinese market, and products in other markets."

Consumers swarmed to the Excelle when it was put on the market at a price of just over 100,000 yuan in 2003. Others were cutting prices, but that didn't deter them from raising the price of the new star.

When the new car was released, I remember Shanghai GM General Manager Chen Hong saying, "We are standing on the shoulders of giants. Now we must become giants." That's how ambitious they were.

Confronted by the strength of the A-class car manufacturers, an indirect strategy of introducing lower-grade A0-class small cars was adopted. Shanghai GM's Sail, Shanghai Volkswagen's Polo, Nanjing Fiat's Palio, and Tianjin's Xiali 2000, all launched in 2002, creating waves in the car market. They became known metaphorically as the Four Flowers, after young female roles in Peking Opera.

The Four Flowers were the first models introduced in Chinese automobile history bought by ordinary people. They brought about a large-scale shift from the use of public funds to household consumption for automobile purchases. They comprised a 100,000 yuan price class, and made it possible for more working-class people to get a family car, something that was just a dream ten years earlier.

But the media and consumers forget that the Four Flowers pursued the world's highest technical standards. Shanghai Volkswagen led the competition in this regard. The fourth generation Polo was launched only six months after its launch in Germany. New technologies were adopted, with changes in components and standards carried out in a dynamic process. Technical operations and management were much more difficult than people thought. The third generation Polo could have been introduced with existing technology and second-hand equipment, saving a lot of labor and keeping it price competitive. Yet, they chose the more difficult path.

To be a front-runner, you have to invest huge financial resources and energy. To produce the fourth generation Polo, Shanghai Volkswagen invested 3.4 billion yuan to build a new factory with flexible production technology, which was capable of mixing and producing various models of the Polo series. Laser welding and welding robots were used to ensure accuracy. The body was made of double-sided galvanized steel, and cavity wax injection technology was used to ensure anti-corrosion of the vehicle for 12 years. These investments, invisible to consumers, brought them more advanced technology.

Family cars were for the general public, so the market had to be diverse. The Geely, Xiali, and the Chery QQ, at 30,000 to 50,000 yuan, all met national safety and environmental standards and had enthusiastic customers. As leaders in advanced technology, the Four Flowers let Chinese people drive the same small cars that Americans, Germans, and Japanese drove, just as they used the same digital TVs and broadband internet.

Mercedes-Benz perseveres for 20 years in Beijing

DaimlerChrysler and Beijing Automotive Holdings reached an agreement with a total investment of 1 billion euros in September 2003 to jointly produce Mercedes-Benz cars and trucks. Mercedes-

Benz had been trying to enter the Chinese industry for nearly 20 years.

In the 1980s, Mercedes-Benz transferred technology to China's weapons industry to produce heavy trucks in Inner Mongolia, pioneering the production of high-grade, large-tonnage trucks in China. After that, FAW assembled eight hundred and ninety Mercedes-Benz cars for use by high level officials, but it wasn't followed up. China's Automotive Industry Production Policy was announced in 1994. Mercedes-Benz was compelled to use the advanced A-class vehicle technology it was developing to produce a small car as its Family Car China (FCC) concept, but it had difficulty gaining a foothold. The already approved Mercedes-Benz joint venture van project in southern China also ended in discord. Finally, only a joint venture bus project in Yangzhou came to fruition. Mercedes Benz persisted, but it was disheartening for them.

Daimler merged with Chrysler, the third largest automobile company in the US, to form the DaimlerChrysler Group in the late 1990s. The ambitious DaimlerChrysler had at one time discussed producing Mercedes-Benz commercial vehicles with FAW. However, the two sides came to an impasse when FAW General Manager Zhu Yanfeng insisted on retaining FAW's own Jiefang brand. DaimlerChrysler then turned to Beijing Jeep as a way of entering the Chinese auto industry by introducing its Mitsubishi SUV, but it still hesitated to produce Mercedes-Benz cars there.

In a turn of events, DaimlerChrysler decided to cooperate with BAIC Group to establish Beijing Benz Daimler-Chrysler (BBDC) in 2003. There were three compelling factors. Firstly, there was a surge in the Chinese car market; secondly, BAIC had emerged as a strong and reliable partner through its joint venture with South Korea's Hyundai; and thirdly, the market share of Mercedes Benz's imported cars was directly threatened after its old rival BMW began rolling out its new 3 Series and 5 Series cars in a joint venture with Brilliance.

BBDC decided to stop production of the Jeep, and build a new plant in Yizhuang, Beijing to produce Mercedes and Chrysler cars. The domestically produced Mercedes-Benz E class was unveiled in the Yizhuang Development Zone in 2005. It was basically assembled from imported SKD parts. Because of preferential taxation for local production, it was affordable and sales were brisk.

China's Management Measures for the Importation of Vehicle Parts and Components came out at that time. The previously mandated 40 percent localization for joint venture products was changed to 2+3, that is, the two major components (engine and body assembly) plus three minor components. Alternatively, it could be one major and five minor or 1+5. This threshold was obviously an effective constraint on importing whole vehicles and only putting on the tires in China to obtain a preferential local tax rate.

Achieving a 2+3 or 1+5 localization with a high-end car on a small-scale of only 20,000 vehicles was easier said than done. It wouldn't be worth the huge investment in building a factory in China. Locally produced Audi and BMW parts were already well established, making domestically produced Audis and BMWs more competitive. Mercedes was in a bind.

However, as only Mercedes-Benz could, it set up a China sales company to strengthen its brand and expand imports of all of its models. Meanwhile, it merged existing Chrysler production capacity with the Beijing Benz factory and began improving the technical and quality standards of its local component suppliers. This created the conditions for domestic production of C-class vehicles and new long wheelbase E-class vehicles.

Interestingly, Europe, the US, Canada and other countries later sued China in the WTO because tax breaks for localized parts violated import rules. The lawsuit played out over three years. The complainants finally won, and China had to revoke its Measures for the Administration of Imports of Vehicle Parts and Components. But Beijing Benz, the primary beneficiary of the lawsuit, had already reached the requirements for localization and had benefited from reduced costs. These were the games countries played when China joined the WTO.

The new Mercedes-Benz A class was launched on the luxury yacht Aida in the Aegean Sea in July 2004.

2. Competition becomes bloody

The good old days come to an end

Although media and consumers poured into the pavilions at the 2004 Beijing International Auto Show, the Chinese auto market, which had maintained growth of more than 50 percent for two years running, suffered a 20 percent drop in sales in May 2004. More than 20 percent of total output was stockpiled.

Whether it was a consolidation of the bull market or a turning point, one thing was certain, the good old days, when any car could be sold and money made, were gone forever.

Chinese automobile culture had been held back a hundred years and there was pent up demand. Sixty percent of the first people who bought cars didn't even open the front hood to see what type of engine it had.

In that explosive market environment, manufacturers enjoyed almost unlimited growth. Everything sold well, from the Geely and Xiali at 30,000 to 50,000 yuan to the record setting 10 million yuan prices of Bentleys and Rolls-Royces. The joke going around was that at least you could sell cars even if you couldn't do anything else.

For international automakers who were experiencing a global downturn, this was an opportunity to force their way into the Chinese market. And it gave them extraordinary returns too. Foreign media and investment analysts thought it was insanity when Shanghai GM and Guangzhou Honda began planning and building their factories. Now they were the money trees for their parent companies. Volkswagen Group, the earliest to enter China, got half of its profit and output from China. However, the market stopped growing, incremental growth ended, and there was no room for newcomers.

When the auto market began to panic, Volkswagen suddenly dropped a bombshell on the night of 16 June. Volkswagen China joined by its two joint venture partners, FAW Volkswagen and Shanghai Volkswagen, lowered prices on all of its products. A vicious price war started, and competition became bloody.

Multinational companies still see China as a tasty bite

People were caught off guard when production and sales growth fell after May 2004. However, some multinationals weren't put off by the downturn. It was still a tasty bite, despite the increased competition.

Carlos Ghosn, CEO of Nissan Motor Co., visited Beijing on 20 July. He was asked by a Chinese reporter if Dongfeng Nissan would cut production over the next four years, given the fall in the market. He answered calmly, "In the first half of the year, the US auto market grew by 1 percent, Europe's growth was zero, Japan's was a negative 4 percent, while China grew by 21 percent. Isn't it obvious? China's auto industry achieved 50 percent growth for two years running, but that cannot be sustained. Never mind 20 percent growth, even 10 percent gets people excited. We plan on average annual growth of 10 percent, so we won't be making any changes."

Ghosn, who was famous for bringing Nissan back to life, used to travel around China in a chartered aircraft. In a single day, he inspected production lines and discussed plans in Huadu in Guangzhou Province, then in Xiangfan and Shiyan in Hubei Province, and flew to Beijing overnight. The following day, he met Miao Wei, Chairman of Dongfeng Co., Ltd., went to a dealership for a working lunch with employees, and was interviewed by the media. That evening, he attended a launch event for the Scorpio luxury car. His schedule was breathtaking.

Jim Padilla, Chief Operating Officer of Ford Motor Company, appeared on the same day as Ghosn. He rushed from Beijing to Nanjing for the signing ceremony of Chang'an Ford's factory in Jiangning. This was the second new 200,000 vehicle factory built by Ford in Shunjiang since the successful launch of the Mondeo sedan in Chongqing. Japan's Mazda Motor Corporation President, Ichikawa, also showed up at the ceremony. Since Mazda was a subsidiary of Ford at the time, the new Nanjing plant was expected to produce the Mazda3 and the Ford Focus, developed by Ford for Asia, on the same chassis.

When Ford developed the Focus sedan for the Chinese market in 2005, Chinese media were invited to Europe to observe the R&D and testing.

Ford's Pan-Yangtze River plan surfaced, almost completely escaping the media's attention. Intellectual property rights and model adjustments had been secured by Ford's Jiangling Motors Holding Company in Nanchang on the middle reaches of the Yangtze River. The move was the centerpiece of Ford Chairman Bill Ford's announcement in October 2003 of a US$1 billion investment to expand in China. It was a rapid response to GM's recent announcement that it would increase its investment there.

Ten days earlier on 10 July, VW's China Affairs Vice President, Wei Zhibo, flew to Changchun. Together with the General Manager of FAW Group, Zhu Yanfeng, they laid the foundation stone for Volkswagen FAW Chassis Components Co., Ltd. It was the first advanced chassis production plant established after Volkswagen came to China. The main products for this 1.42 billion yuan investment were the principal components of the PQ35, Volkswagen's new generation A-class chassis. These included the suspension, steering, and braking systems. VW's most advanced models, the fifth-generation Bora (Sagitar), Audi A3, New Beetle, Touran and Kaidi, could all be assembled with these components. This significantly increased the competitiveness of Volkswagen's two joint ventures in China.

The introduction of the most advanced PQ35 platform ended a 20 year history of obtaining parts certification from Wolfsburg. The chassis plant was expected to be commissioned two years later with an annual production capacity of 400,000 units. Volkswagen and FAW jointly invested in the construction of a new engine plant in Dalian to produce advanced turbocharged engines with small displacement and low fuel consumption. After 20 good years and faced with hostile new competitors, the Germans finally realized that if they merely stuck to the rules, VWs share of the Chinese market might not be as good as before.

New leaders and new policies

There was a generational change in the leadership of Chinese auto companies around 2004. The top leaders of the three major groups changed from Yan Zhaojie to Yan Yanfeng at FAW, from Chen Qingtai to Miao Wei at Dongfeng, and from Lu Ji'an to Hu Maoyuan at SAIC.

They were all under 40, and were called the Three Young Marshals[3]. Leadership changes also took place at other state-owned groups with Yin Tongyao at Chery, Yin Jiaxu at Chang'an, Zeng Qinghong at Guangzhou Automobile, and Xu Heyi at BAIC.

Joining the WTO and encouraging family car ownership accelerated structural adjustment in China's auto industry. It was transformed from a living fossil of the planned economy to a competitive market entity. The previous Automotive Industry Production Policy of ten years had clearly failed to keep up. Keeping pace with the times, the just established National Development and Reform Commission (NDRC) issued a new Automotive Industry Development Policy on 1 June 2004.

This policy eliminated inconsistencies between the WTO rules and commitments made by China by abolishing foreign exchange balances, localization ratios, export performance, and other requirements. Administrative requirements were reduced substantially, using regulations guided by technical standards instead. For the first time, a brand strategy was proposed and enterprises were encouraged to develop products with their own intellectual property rights. Mergers and acquisitions were put forward, encouraging enterprise groups to become bigger and stronger. Enterprises were required to overhaul their sales methods. Development of energy saving and alternative fuel vehicles was also encouraged.

These provisions might seem somewhat general and dull to the layman, but the principal direction and major events in the development of the industry followed this general framework. A good city plan doesn't need to be changed very often, just as Chicago is still

---

[3] Young Marshal is a reference to Zhang Xueliang who was the effective leader of Northeast China and much of northern China from 1928 until the Japanese invasion in 1937. He was an instigator of the so-called Xi'an Incident in which the Chinese Nationalist Party leader Chiang Kai-shek was kidnapped in order to force him to enter into a truce with the insurgent Communist Party and form a united front against the Japanese who had occupied Manchuria. After the Xi'an Incident, Zhang was placed under house arrest by Chiang for the next 50 years, first on the mainland and then in Taiwan. Zhang was considered a patriotic hero by the Communist Party for his role in the Xi'an Incident.

rigorously implementing its vintage plan for urban development from 50 years ago.

Output grew by nearly 20 percent, but growth of sedans was only 13 percent in 2004. Although this was the best in the world, most car manufacturers, dealers, and component suppliers felt otherwise.

More than half of the grace period since joining the WTO had passed. However, the harsh reality was that it was easy to lower prices but hard to reduce costs. Although China was 80 percent cheaper than Japan and other countries in terms of labor costs, its procurement, production, and management costs were higher. Costs for joint ventures in China were 20 to 30 percent above the international average. Procurement costs of parts and components were 50 percent higher, manufacturing costs were double, and daily operating costs were eight times the international average!

Furthermore, fuel provision, roads, and parking capacity were overwhelmed and couldn't accommodate the sudden increase in cars. Why would anyone buy a car if the roads were like parking lots?

When an experienced farmer plants wheat, he doesn't just let it grow wildly with the arrival of spring. A period of restrained growth is also needed to make the seedlings healthy and strong. After the auto market passed this germination stage, savvy car manufacturers turned from increasing capacity and prices to paying attention to markets and reducing costs to survive. The good days of competitive growth were gone and would never return. Competition became fierce. In this survival of the fittest, companies came and went, and competition became the norm.

Annual output of cars was 2.76 million in 2005, an increase of 20 percent. It was 3.87 million in 2006, an increase of 40 percent. The Chinese auto industry was safely through the grace period and on a firm footing.

04 A good man: An impression of Lü Fuyuan

A friend told me on the phone that Minister Lü was gone.

I was speechless for a while. I knew he was ill and had expected this day to come.

As a reporter who has covered the macro-economy for many years, I have a special fondness for the automobile industry. One reason is that many of my friends in it are fine people who are knowledgeable, open-minded, and easygoing. Lü Fuyuan was one of them. Whenever he was mentioned, I always replied that he was a good man.

A grassroots minister

When I met Lü Fuyuan in the early 1990s, he had just transferred to the reorganized China National Automotive Industry Corporation as Deputy General Manager. When I interviewed him, or when we went on business trips together, we got along like friends. He wasn't the slightest bit official or formal, rather he was refined and somewhat bookish.

When we talked about cars, his vision was broad, and his views were unique. He put Chinese cars into an international context, which was valuable in those relatively closed years. I used to joke that I needed to bring a notebook when I chatted with him. For me as a journalist, there was so much in what he said that if I missed anything, it would be my loss.

He was a grassroots minister. When he graduated from the Physics Department at Jilin University during the Cultural Revolution, he was assigned as an electrician in a small town in Lishu County in Jilin Province, and got into the FAW Hongqi Car Factory by swapping with someone. He started as an air-conditioning assembly worker installed and calibrated FAW's first three computers through his own reading and research, and learned programming. He went from worker to technician to engineer, and was sent to the University of Montreal in Canada as an exchange scholar in 1982. When he returned to China, he was the Deputy Head of the Technology Department, Deputy Factory Director, and Chief Economist. He hosted important negotiations on innovating and bringing technology to FAW. He was fluent in English and was a quick thinker. He was recognized as an expert negotiator and was known as FAW's Kissinger. As an industry leader, his strategic vision, overall sense, and meticulous methods were soon recognized by auto manufacturers everywhere.

A bookworm

Not many Chinese officials make hanging around in a library their holiday pastime, but Lü Fuyuan never tired of it. There was no internet then, so he would spend whole days at the National Library reading up on the latest information about the global automobile industry. He said a good thing about working in Beijing was that it had the biggest library in China.

His wife, Teacher Miao, told me that on the first Sunday when the family moved to Beijing, the couple hurried straight to the National Library with their children, carrying bread rolls and a flask of water. Lü Fuyuan spent the whole day in the Chinese and foreign language reading room with their son. Their daughter was too young to get a library card, so she learned how to search the catalog with her mother. When Lü Fuyuan went abroad on official business, his favorite hobby was strolling around second-hand bookshops. Most people spent all the foreign currency they were issued on color TVs and refrigerators, but Lü Fuyuan used all his to buy books.

I, like many people, love to read, but I'm not up to Lü Fuyuan's level. Once when he was overseas, he spent a whole day reading in a second-hand bookshop. As he was settling the bill, the proprietor said, "I enjoy guessing my customers' professions by the books they choose, but you have read widely and I can't guess what you do." Lü Fuyuan said that he was a teacher. The proprietor said, "I've got an out-of-print book here. If you can tell me what it's about, I'll give you a fifty percent discount on all the books you've bought today." He brought out the book, and was very impressed when Lü Fuyuan told him not only what was in it, but also some of the background to it. It was about global economic trends after World War II and was very insightful, but the author died early, a small number of copies were printed, and only a few still existed. Lü Fuyuan had read a review of the book and later found it in a library in Britain. He was fascinated by it and made a photocopy to take back to study in China.

Lü Fuyuan moved several times, from Changchun to Beijing, and from engineer to minister. In his living room, there was always a simple, imitation leather sofa made at the FAW factory in the 1980s. The other furniture was a few large bookcases filled with books.

Cars are productive too

It's not easy to be an official without saying things you don't want to. In the mid-1990s, a department held a media briefing, trying to put the brakes on a "car fever" that really did not exist at that time. Lü Fuyuan, Vice Minister of the Ministry of Machinery Industry in charge of the automobile industry, was invited to speak. He read for a while from his prepared script, then put it aside and talked about his own ideas of how the automobile industry would promote growth in the national economy. I have quoted what he famously said in that speech many times: "This country of 1.2 billion people is modernizing, that's why we have to develop cars to carry people. The flow of people is more important than the flow of goods. A car carrying four people is more productive than a truck carrying four tons of rocks."

Car ownership in Beijing surged to more than 800,000 in 1997, traffic jams appeared, and a clamor arose to limit the number of cars.

I had been calling for households to have cars for many years. I asked those who wanted to restrict them what a reasonable number was, or if they had a number in mind. None of them could give an exact answer. Lü Fuyuan, was the first to put forward a planned total number of motor vehicles in Beijing. He pointed out that more problems could occur if Beijing did not plan and build for car ownership of 3 to 4 million vehicles. In major international capitals which are economic, political, and cultural centers, car ownership is at least 4 million, and as much as 7 million. And they have developed underground mass transit. The overall size of a city's economy is always linked to its car ownership. If there's not enough, the economy can't be fully active.

Still a car man

Lü Fuyuan was close and candid with his friends. In 1998, before I finished writing *The Lure of Family Owned Cars in China*, I arranged for an extended interview with him about long-term growth in the automobile industry. I made him wait two hours because of an unexpected traffic jam. I was embarrassed when I arrived at the Ministry, but he sat me down with a minimum of formality and began to speak. As an activity with the Electronics Ministry Party Group was scheduled, he just gave me an overview of the points he had prepared.

Later, he became a Vice Minister of Education and no longer publicly commented on cars. We met sometimes and he said being involved with education was very interesting and he was throwing himself into it. He patiently answered my questions about the car industry, but made it clear he didn't want to be quoted.

I remember the date of our last in-depth conversation. It was 2 January 2001. I ran into him unexpectedly on a plane back to Beijing from Australia.

The long trip was ideal for a conversation. We talked about the slow pace of construction of the country's highway network. Lü said that it was because the cities were keen to build small rings around themselves, and weren't giving enough thought to major connecting roads. He said that Beijing had to build satellite cities to unburden itself, and improve environmental and living conditions. By building large malls in satellite cities, bringing in brand-name chain stores, and creating the same shopping environment as in the city center, people could be enticed to live there and traffic congestion would be cut.

On Brilliance Auto's[4] indigenous development of the Zhonghua, he said it didn't matter if there were problems with the new car. There had been major problems with the Mercedes-Benz A series, but they turned it around. "The key thing is to have mechanisms for solving problems. They must focus on operations, but they have to have technical people in charge at the highest level, and a team to deal with manufacturing and technical issues. They can even get foreigners to do it. And they run a great risk without an authoritative chief engineer."

He spoke to me about the globalization that people were so concerned about. He talked about China's need for long-term goals and what the concept of a big market was. He said that the US was a big market, where replacing 5 percent of the cars is 2 million vehicles. China is the only big market that can compete with the US globally in the 21st century.

I recorded the conversation in my diary at the time. Years have gone by, but his insights still resonate.

At the 2003 National People's Congress, Lü Fuyuan was appointed as the first Minister of the newly established Ministry of Commerce. In September that year, he was diagnosed with cancer and urgently needed surgery, but he led a Chinese delegation to the Fifth WTO Ministerial Conference in Cancun, Mexico, the first time for China as a formal member. He took part in 25 bilateral and multilateral meetings and consultations while he was there. The day after he got back to China, he flew off again to Cambodia. There was more than a month of delay before he was hospitalized. However, the best time for surgery had been missed. Lü Fuyuan passed away on 18 May 2004, at the age of 59.

---

[4] Brilliance Auto Group is a Chinese automobile manufacturer headquartered in Shenyang. Its products include automobiles, microvans, and automotive components. Its principal activity is the design, development, manufacture, and sale of passenger cars sold under the Brilliance brand.

The last time I saw him was at a signing ceremony for a strategic cooperation between a domestic automobile group and a multinational company. He asked me why he hadn't heard from me for so long. I said that he had moved, and I couldn't get his home number from the Ministry of Education. He took my notebook straight away, wrote down his mobile number and home phone number, and wrote his name in large characters beside them.

I hadn't thought this outstanding Chinese automobile man would leave us so soon, when the ink was barely dry.

(The first draft of this was written the day after his death. I got the bad news early in the morning, jumped out of bed, wrote it at one go squatting in front of the computer, and put it online straight away. It wasn't until I finished that I realized I had been walking barefoot.)

Chapter 7

Indigenous Brands Become National Policy

Until 2004, officials and the media paid scant attention to indigenous brands.

On February 20th, just as the Two Sessions — the National People's Congress (NPC) and the Chinese People's Political Consultative Conference (CPPCC) held annually — were about to start, Chery held a press conference in the Beijing political conference venue to announce that 600 Chery Eastar (or Oriental Sun) vehicles had obtained qualification from the NPC. Two days later, a female Xinhua reporter published a report in which Zhan Xialai, Wuhu Municipal Party Committee Secretary and Chairman of Chery, was called the Top Red Businessman. It caused a political storm. A veteran CPPCC comrade declared, "We won't drive the Top Red Businessman's cars." Taxi companies immediately replaced their Eastars with Red Flags. Zhan Xialai resigned from the post of Chery Chairman and it was run by Yin Tongyao. You can imagine the pressure on Chery.

However, it turned out well. That summer, central decision-makers began to support indigenous brands and innovation in the auto industry. Chery, an indigenous brand state-owned enterprise, led a charmed life for a time. As it entered a period of great vigor, it was overwhelmed with leadership inspections, high-end forums, and policy support. But the major state-owned joint ventures were brought to ruin one after the other.

It's easy to show favoritism in China.

1. Support for indigenous innovation at decision-making levels

Indigenous bottom feeders liberate themselves

From 2004, policymakers began raising innovation concepts to a pivotal position, and frequently proposed improving capacity for indigenous innovation as a central link to promote structural adjustment, transformation of growth modes, and national competitiveness. Indigenous brands became famous as their reputation spread.

Indigenous brands, the bottom feeders that had been almost killed off in the previous few years, suddenly became attractive, drawing official and media attention. Two or three years earlier, Chery, Geely, Brilliance, and Yueda had been applying to government departments for the right to exist. Now, honors, policy support, and even loans and special funding rained down on Chery and the others.

I always thought it strange that the indigenous brands hadn't encountered those favorable officials and media in the past two years. Of course, revolutions occur in no particular order, especially when reforms reach a new stage and policymakers determine a new course.

There were two generations of indigenous brand cars which seemed unrelated to each other.

In the heat of the Great Leap Forward in 1958 and at a time of international embargo, FAW's Dongfeng and Red Flag, Beijing's Jinggangshan, and Shanghai's Phoenix relied on copying imported prototypes, fabricating them by hand. Those cars were not commodities, but a manifestation of the political enthusiasm of the time. When the Great Leap Forward bubble burst, there were only two brands left. One was the Red Flag, with annual production of no more than 300. The other was the Shanghai brand which was renamed Phoenix, with annual production of 3,000. They carried on for more than 20 years. It was a pity that these two first-generation indigenous brands went out of production when large numbers of imported cars poured into the country as we entered the 1980s.

Second-generation indigenous brands like Geely, Chery, and Brilliance began popping up at the end of the 1990s, all privately run or state-owned enterprises outside the automobile industry. Their biggest difficulties in setting up were the high threshold of industry access and official indifference. It wasn't until we joined the WTO, and multinational companies could enter joint ventures, that indigenous brands got the right to exist.

Cars made by Chinese people went through three stages of development, from self-reliance in 1958 to the introduction of joint ventures in the 1980s to indigenous development and indigenous brands after joining the WTO. Indigenous development was thought to be empty talk in the mid-1980s, when there was no investment, talent, technology, or markets. Development was finally possible in the late 1990s, when the introduction of joint ventures brought an accumulation of talent, technology, and components. Brilliance developed Chinese cars with global resources, and Chery and Geely launched good cars that Chinese could afford. Unfortunately, they met with suspicion and blame. By 2004, Chinese car production exceeded two million. Driven by the policy-makers, government departments began to support and encourage indigenous brands. The fruit was finally ripening, even though it was late.

Hard-won opportunities

The Chinese car industry really began with the introduction of joint ventures. It had its own historical context and the successes were hard to deny. However, some government departments deviated from the original idea of obtaining advanced international technology and forming indigenous innovation capabilities. They even suppressed the new brands. That was not the intention of the policy makers and creators of joint ventures.

Geng Zhaojie, founder of FAW-Volkswagen and former FAW Factory Director, once told me that the aim of introducing Audi technology and establishing a joint venture was to learn advanced global auto technology to create a second-generation Red Flag, which FAW people had dreamed of doing for decades. That was difficult for outsiders to understand. He stressed that the joint venture and the indigenous brand had to be like two legs of the same thickness. It was not their intention for one to be thick and one thin.

Chen Hong, General Manager of SAIC, said, "If we can't establish ourselves through indigenous innovation in the next 20 years, others will have to help us do it. To be a manufacturing industry, we have to get to the core of the value chain and strengthen our technological R&D capabilities. Fully competing isn't possible at the beginning, but partial breakthroughs can be made. The indigenous innovation and brands we are engaged in now are different from those in the 1960s and are carried out under conditions of openness. We are talking about cooperation, but we won't give up the element of independence."

Central policymakers raised indigenous innovation to the level of national policy in 2004. Indigenous brands finally emerged from the awkward situation of being loved by no one, standing tall at last. They cherished their hard-won opportunities, fought hard, and made breakthroughs.

A number of grassroots indigenous brands such as Chery, Geely, Brilliance, Chang'an, Great Wall and JAC[1] were rising fast. There were praiseworthy achievements in brand building, R&D, product export, and overseas construction. Lifan and BYD entered the car industry and gained a foothold in the fierce market competition with unique core competitiveness. Indigenous Chinese brands began to emerge, and their booths were visited by multinational CEOs at international auto shows in Detroit, Frankfurt, and Tokyo.

The high-profile advocacy of indigenous brands by the Central Government is worth mentioning. Many state-owned automobile groups made full use of the funds and technology introduced into joint ventures over the years, and carried out R&D at the group level to create third generation indigenous brands, mainly mid- and high-end models. SAIC bought into Korea's Daewoo Group, took over Korea's Ssangyong, acquired British Rover technology, and established its own

---

[1] JAC Motors is a Chinese state-owned automobile and commercial vehicle manufacturer. The company is based in Hefei, Anhui Province, China.

Roewe brand. FAW and Nanjing Automobile also launched mid- and high-end products such as the Besturn and the MG.

With China's large population and geographical extremes, consumption is more diversified than in Europe. A number of market segments enabled indigenous brands to gain valuable space in competition with multinational companies.

Indigenous brands made up nearly 40 percent of the total 5 million domestic passenger cars sold in 2007, nearly dominating the export market.

Indigenous brands began to enjoy policy support from government departments. The State made preferential policies for foreign investment and joint ventures at the early stage of Reform to attract foreign capital. In the automotive industry, average tax was 11 percent for joint ventures, 22 percent for domestic private enterprises, and 30 percent for state-owned enterprises. Adapting to changes, China ended supernational treatment for foreign capital and joint ventures in 2007, and car companies were able to compete on a fair tax platform.

Markets have no sympathy for weakness. Competition will decide whether China's indigenous brands finally take their place in the fierce global market. Self-improvement is the key to that. They can only win with their own core competitiveness, product quality, and good service.

2007 became a peak year for indigenous Chinese brands.

2. Indigenous research and development is critical

Chery: Not yet a stalemate

In 2007, Chery became a successful example of the indigenous innovation pathway proposed at the highest policy-making levels. Zhan Xialai and Yin Tongyao were chosen as representatives at the 17th Party Congress. During the congress, I spoke in depth with them about indigenous brands.

"China is becoming the world's largest auto market. It's the last big market for all of the multinational companies. There are good opportunities for indigenous brands, and global competition is right at their front door," Zhan Xialai said. "From a cultural perspective, the Germanic and Yamato races both have a world domination complex. Economically and industrially they strive to be first, and that hasn't changed. When you deal with Germans and Japanese, whatever they say even about minor products, their overall goal is to come first."

How do Volkswagen and Toyota compete to be first in the Chinese market? "With new technology. As a rule, the more intense the market competition, the faster the entry of new technology. Volkswagen had no rival in China for the first ten years, but when Santana had been going for 20 years and competition was in full swing, the advanced FSI engine, the dual-clutch gearbox, and the PQ35 and PQ46 platforms all came in. Toyota also brought the Camry and New Corolla to China. It was competition that made them do it," Zhan Xialai said.

China had become a most competitive part of the international market, and the situation with indigenous brands was quite grim. Zhan Xialai said, "It's clear that indigenous Chinese brands are surviving in the market gaps. They are competitive, and they have advantages with labor costs and efficiency, but price wars can't be their only strategy. To persist long term, they have to align product development to quality service, and perform as well as the multinational companies. Indigenous brands will have to conserve energy and be tactical for some time to come. They have to determine the connection between a product and a series and lay out a unified strategy. It is a protracted war, not a matter of days or even years."

Chery's R&D spending made up 7 percent of total sales revenue in 2005. Yin Tongyao, Chery Chairman and General Manager, believed it was that 7 percent investment which enabled Chery to develop new engines, two families of gasoline engines, and diesel engines. It was a strategic turning point when Chery's one millionth car came off the assembly line on 16 August 2007.

Innovation to create an indigenous brand had been the first stage of Chery's development. Chery brand cars were exported to 56 countries and regions and Chery won the China Auto Export Championship three years running. As the Chery people would say, "Let's get on with it."

Chery then made its move and began the second phase of development, sticking with innovation and building an indigenous international brand. At first it was under the banner of an indigenous brand, but they backtracked to form a joint venture with Jaguar Land Rover. Yin Tongyao said Chery wasn't satisfied being a domestic or regional brand. From then on, they would go out and be part of the global value chain and become a world player.

Zhan Xialai compared it to Mao Zedong's comment in his book *On Protracted War* — they hadn't reached a stalemate in the eight-year war of resistance against Japan and they were still not as strong as the enemy.

Geely: Making an appearance at overseas auto shows

Cars had been driven around the world for a hundred years by the time Chinese people started making them. It was really the fearlessness of the ignorant at first. In Li Shufu's words, they were just "two sofas with four wheels." Imitation was unavoidable. When you learn to write with a brush, you have to start by tracing examples. It was understandable that imitation had to be relied on to begin with.

I was at Ningbo in the summer of 2005, where Li Shufu, the Chairman of Geely Automobile, told me he was wondering whether to go to the Frankfurt Motor Show. I said straightaway, "Yes! It's a good chance for the Geely brand to make an appearance. An indigenous brand taking part in a first-class international auto show for the first time will make headlines. All the Xinhua News Agency and CCTV reporters in Germany will be interviewing you, not to mention the foreign media."

Li Shufu had just created the Free Cruiser, a completely indigenously-developed new car. It was all Geely's own, from engine and automatic transmission to chassis. He wanted his creation to appear at an overseas auto show to show car bosses around the world that China Geely had said goodbye to imitation. He had realized that active development and innovation were necessary conditions for survival and competitiveness.

Their grassroots origins always dogged Geely and Li Shufu. Chinese people's concept of status led officials, colleagues, media, and even consumers to place Geely at the bottom of the Chinese auto industry. And multinational companies were wary of this privately operated enterprise. But once Geely was discovered to be a potential competitor, the blows were often unkind.

When production of the Geely HQ started in 2001, it was fitted with Tianjin Toyota's 8A engine. Toyota's small cars in China were not selling in large numbers. They could make plenty of engines and the contract signed with Geely was generous. Each engine was priced at 17,600 yuan, which could be discounted if they bought more later. However, Toyota breached this contract when Geely's production increased sharply, competing with its small car. Each engine rose to 21,500 yuan, the supply was strictly controlled, and quality compensation clauses were no longer upheld. Geely, realizing that the heart of their product was in the hands of outsiders, immediately began developing its own engine. Geely's MR479Q engine came out a year later, at half the price of Toyota's. "What's the technical standard like?" I asked Li Shufu. "It's just as good as the 8A!" he said. Geely kept on developing the engine. The 2005 1.8-liter engine, with an output of 57.2 kW, was as good as the world's best.

There was nothing Li Shufu would not do. Multinational companies had a tight hold on automatic transmissions. The State, intending to develop automatic transmissions locally, had arranged nationally-funded projects for SAIC and Tianjin Auto. They went on for two or three years, investing more than eight hundred million yuan and involving hundreds of experts. Around 2002 the money had all been spent, but the project was unfinished. It was too difficult. They weren't as good as imports, so it stopped. Li Shufu, who had started out recovering precious metals from waste, decided there was gold to be found in the failure, and it would be a pity to give it up when eight hundred million yuan had already been spent. He went to an expert on the research team, Xu Bingkuan, Chief Engineer of Tianjin Gearbox Factory, and asked him if they keep going, could it be done? Xu said that there was a 50/50 chance. Li Shufu asked him to establish an automatic transmission company at Geely. The inspection lines were designed by Geely. The high-precision probes, inspection instruments, precision machine tools, and processing centers were imported from Germany and Japan.

To import equipment from Japan, they had to go there to examine and take charge of it. However, Li Shufu and Xu Bingkuan kept being refused visas, so they couldn't go, and their equipment couldn't be brought in. "Not at all open-minded," Li Shufu commented. In 2005, three- and four-speed automatic transmissions were built and installed in the new Free Cruiser. Li Shufu said that they still weren't finished yet. Ultimately, Geely's acquisition of DSI Australia, the world's second largest independent automatic transmission plant, had everything to do with its understanding of automatic transmission technology.

In 2005, Li Shufu announced that Geely Automobile's annual sales would reach 2 million units by 2015, and two-thirds would be exported or produced overseas. The acquisition of Volvo hadn't even been thought about yet. That year, Geely sold only 140,000 low-end cars such as the Merrie, HQ, and Uliou, and exported 4,846 vehicles. It was hard to convince most people of the need to surpass this growth rate. Since the overseas market would make up two-thirds of Geely's sales in ten years, it was logical to make a showing at world-class auto shows like Frankfurt.

A private Chinese company going to the auto show at Frankfurt was something even the Germans weren't mentally prepared for. Almost half of the attendees, including Li Shufu himself, were refused visas by the German Embassy. Li Shufu applied again but was only given a six-day stay. He didn't take offence; to the Germans it made sense. The Chinese car went, and Geely represented China with the five-star red flag. When it was flown at the Frankfurt Motor Show for the first time, Li Shufu gained more than just national pride.

I went to see Geely as soon as I got to the Frankfurt Motor Show in September 2005. After searching for a while, I finally saw its booth in a corner of a non-mainstream brand pavilion. It was quite a large area, and the backboards were painted with large peonies in a Chinese design. There were five cars on the stage including the HQ 203A right-hand drive export car; Geely's second-generation sports car, the Chinese Dragon; China's first automatic transmission Free Cruiser; and the newly developed Geely top model, the FC-1. Instead of beautiful fashion models presenting the cars, there were Beijing opera performers in full costume. This attracted a lot of attention. Geely's press conference was scheduled on the afternoon of the second day. Li Shufu personally drew away the red flag covering the Chinese Dragon as cameras flashed. I noticed that world-class car magnates like GM CEO Rick Wagoner seldom entered that pavilion, but this time he took a team for a special visit to the Geely booth.

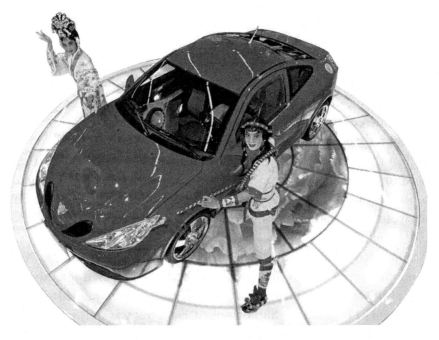

The Geely, in its booth at the Frankfurt Motor Show, was the first Chinese indigenous brand to attend an A-class international auto show. Li Shufu displayed the car with Chinese paintings, peony flowers, and Peking opera characters in the background.

Geely spent 10 million yuan to go to Frankfurt. The world's auto industry learned that the Chinese Dragon had arrived. China had Geely just like India had Tata. Xu Guozhen, then Ford Global Communications Manager, said that Geely would afterwards be able to maintain its international presence.

Big groups: Achieving big things

The car market is global, and the Chinese market is an arena for international competition. For an indigenous grassroots brand fighting alone, it is increasingly difficult to compete and avoid being squeezed out by international brands. Government economic departments directs central enterprises and large state-owned auto groups to create indigenous brands.

Large-scale business groups have a lot of capital and great technical prowess. If you begin establishing indigenous brands starting from the low end, you will encounter bottlenecks. But, if you start from a high point, your product strategy will have clear long-term goals.

The indigenous Pentium brand developed by FAW was based on the Japanese Mazda 6. At a meeting before its launch, I suggested to Zhu Yanfeng, FAW Group General Manager, "What about calling it the Liberation? It was China's first automobile brand, and has been recognized widely for over 50 years." However, friends at FAW didn't agree, saying that Liberation was a truck brand, a lowly start for a car. In fact, famous brands such as Mercedes-Benz, Volvo, and Nissan had all succeeded with commercial vehicles and passenger cars coexisting under the same brand. Many years later, then FAW General Manager Zhang Pijie, told me my suggestion wouldn't be unreasonable today.

Some people worried that FAW's strategy of using the Mazda 6 platform for the Pentium would make it lose its position on subsequent indigenous R&D platforms, and it would also be difficult to withstand the impact of Mazda's own price cuts. However, using the development capabilities FAW had accumulated for 50 years, Pentium lived up to expectations. The success of the Pentium B70 was followed by the launch of the B50. Where Pentium's two joint venture partners Toyota and Volkswagen excelled was in establishing its reputation among consumers as having come from the Mazda 6 but better.

Zhang Pijie told me that while the Pentium B70 was being developed, altogether 518 bench tests, 3 million kilometers of road tests, and 78 actual vehicle crashes and rollover tests were carried out, and unique human rollovers and static pressure tests were done. The high-rigidity body achieved the national crash test five-star standard. To ensure product quality, Pentium set down 55 management objectives in 8 fields, and did strict quality control to make vehicle review scores meet or exceed the standards of foreign products. FAW also paid attention to personal initiative, advocating quality awareness among all employees. He told me about a young female employee in the Quality Assurance Department who was sent to receive products at a component supplier factory in Ningbo. Her family home was only 30 kilometers away from the factory, and her mother had waited for her to go back for dinner every day, but she worked continuously for 11 days without taking time to go home. She went back to Changchun with the quality assured parts as soon as the work was done.

SAIC, the largest and most technologically and financially powerful state-owned automobile group in China, launched the indigenous brand SAIC Roewe in 2006, in a way totally different from Geely, Chery, and others, which had started from scratch.

Red Flag was the eternal dream for staff of SAIC.

3. Twenty years of loneliness

Indigenous development is the tenth steamed bun

Forged over a hundred years, global automotive technology had matured to a perfect state like China's game of Go. There are nine levels of mastery in Go and a novice who wants to be a winner can't do that in a short time.

Nurturing automobile R&D capability in China, according to FAW General Manager Zhu Yanfeng, was a lonely battle. This statement didn't suit the climate of opinion and was attacked by the media and netizens. However, it was the truth. After ten years of growth and ten years of training, developing ninth level Go skills wasn't easy.

As cars are a global industry, product development is the jewel in the crown.

In 1958, in the face of foreign blockades and in the spirit of self-reliance, Chinese people copied foreign prototypes, hammering out bodies to create the Dongfeng, Red Flag, and Shanghai brands. They were small, technically backward, and unreliable. As soon as the country opened its doors, the market was overwhelmed with imported cars. What we call indigenous development today is nothing like it was then. Indigenous development in the modern sense could only happen when car manufacturing and management technology had been mastered and the basis of a huge component parts system had been established. Indigenous development was just empty talk 30 years ago. People still had doubts when Yang Rong proposed developing Chinese cars in 1998. Only now, when China produces more than 10 million cars, has indigenous development come into its own.

Indigenous development has always been the weak point of the Chinese auto industry. Eager for it to improve, the government and public opinion have chided it for moving too slowly. Chinese people were excited when indigenous cars began appearing at auto shows in Beijing and Shanghai after 2010. But apart from the cheers, people questioned whether it had been wise to bring in joint ventures over the past 20 years. If indigenous development had begun 20 years ago, China would be a big auto developer today.

This made me think of a story. A hungry man got a plateful of steamed buns, ate one but wasn't full, ate two and even nine, but still wasn't full until he had eaten the tenth one. He said crossly that if he had known that the tenth would make him full, he needn't have eaten the other nine.

The development of the automobile industry has its inherent objective laws. China had the same opportunity as Japan and South Korea for indigenous development to go on to dominate the domestic market, but it was held back by the Great Leap Forward and the Cultural Revolution. After that point, with the globalization of the automobile market, it became neither possible nor necessary to do it.

Indigenous Chinese auto research and development is on the agenda today precisely because joint ventures were brought in. They were the first nine steamed buns that were needed.

Recalling the first decades of joint ventures in the 1980s and 1990s, old timers in the automobile industry described it as a tragedy. Reaching the technical standards of multinational companies from such a low base and poor foundations wasn't easy. However, one pinnacle followed another, and the steamed buns added up year by year. Chinese people not only mastered first-class car manufacturing technology, but also created a large, supporting first-class parts industry. And they enjoyed the flavor of local development for the first time.

Those 20 years of joint ventures were indispensable. The talent nurtured and management experience accumulated by Shanghai Volkswagen, FAW-Volkswagen, Shenlong, and other joint-venture car companies, especially the gradual development process, not only made indigenous development of local brands possible today but also greatly benefited later participants. The managerial and technical backbone of local brands like Chery and Geely, as well as world brands such as Shanghai GM and Beijing Hyundai, all had their baptism in the first generation of joint ventures. The huge parts and components system formed in the past 20 years has not only greatly reduced investment and procurement costs for local and multinational brands, but will make the Chinese car industry as competitive globally as the Chinese home appliance and communications industries.

In today's globalized world, it's not worth for our officials and scholars to dwelling on how multinational corporations didn't hand over in-house core technologies and source codes to the Chinese side. When foreigners do business, they are not altruistic like Lei Feng[2]. When they hand something over to you, won't they keep something for themselves? Fairness is a principle of open markets. The fact is that through learning, competition, innovation, and accumulation, the Chinese car industry eventually settled on a system of research and development capabilities.

Strictly speaking, we've just put the tenth steamed bun into our mouths. I met the son of the famous Italian design master Giugiaro in Sanya recently. He said Chinese people didn't have a thorough understanding of car design. They often insisted on too many complicated lines. In fact, a flagship model like the Volkswagen Phaeton only had three simple lines.

With the growth of indigenous brands, and especially when products were being exported, intellectual property rights disputes were bound to become an obstacle. Therefore, we stopped copying. We did what Yang Rong did and paid an Italian design company to do it for us. The indigenous development capability we went on to form has become the mainstream for Chinese cars.

Haipai cars reach maturity

We don't need to be humble about our indigenous R&D capability. Shanghai has had deep automobile roots since economic reform began and we can't underrate the foundations of Haipai cars.

The term Haipai was given to the school of painters represented by Ren Bonian and Wu Changshuo in the late Qing Dynasty. They were

---

[2] Lei Feng (18 December 1940–15 August 1962) was a soldier in the People's Liberation Army and is a communist legend in China. After his death, Lei was characterized as a selfless and modest person devoted to the Communist Party, Mao Zedong, and the people of China. In 1963, he became the subject of a nationwide posthumous propaganda campaign, "Follow the examples of Comrade Lei Feng."

Chinese media test-driving the new Regal on the Rhine.

Shanghai GM participated in all the R&D on the Regal using GM's new platform and turbocharged engine.

The highway along the Rhine is a good place for a test drive.

the originators of market-oriented literati paintings. Later the term grew to generally include Shanghai's local and business culture.

Shanghai embodies a hundred years of Chinese cars.

Old Shanghai was known as an International Car Exhibition, because it used to have more cars than any city in the Far East. It passed on the genes for keeping an open mind, staying fashionable, and achieving skill and quality in cars.

Cars began in Shanghai when a Hungarian named Leinz shipped in two Oldsmobiles in 1901, the 27th year of Emperor Guangxu's reign in the Qing Dynasty. By 1947 there were 268,000 cars in Shanghai, more than five times the number in Beijing in 1949.

The Phoenix, born in Shanghai on 28 September 1958, was the first generation of indigenous cars in New China. It was later renamed the Shanghai and was once the only mass-produced car in China. It was a little known fact that even in the chaos of the Cultural Revolution, a rotary engine, technologically different from international designs, was developed in Shanghai and was installed in a 2-ton vehicle. One hundred fifty were produced.

By 2011, there had been cars in Shanghai for 110 years. A century of history had given a Haipai reputation to Shanghai cars that was open-minded, market-oriented, and trustworthy. They had fine workmanship, high quality, and were eclectic, serving as a bridge between China and the West. Haipai car culture is well entrenched.

This inclusiveness allowed Chinese joint ventures and indigenous brands to thrive in Shanghai's fertile soil, while the pervasive local style stamped its mark on even the strongest partners. Shanghai Volkswagen and Shanghai GM were joint ventures where the Chinese and foreign sides connected the best.

Shanghai Volkswagen's indigenous development took off at the end of the 1990s, creating a precedent for others. In 1997, Shanghai Volkswagen invested more than 3 billion yuan to establish the first technology research and development center, and sent two batches of promising young engineers and technicians to Germany for training. Shanghai Volkswagen took four years to build, and had the largest car test track in Asia. By the turn of the century, it had basically developed the standards and processes of German Volkswagen and the ability to develop interchangeable bodies, electrics, chassis, and transmission systems. The LaVida is a successful model developed by Shanghai Volkswagen in the 21st century. It has maintained the top position in sales of single vehicles for many years. Its design is sleek and symmetrical. It is not a remodeling of a Volkswagen model, nor is it a combination of several models.

In 2015, SAIC and Volkswagen developed the Phideon, a high-end, C-class sedan based on the advanced platform and engine technology of Volkswagen Germany. This model became Volkwagen's first global high-end C-class product.

Shanghai GM, established in 1998, originally focused on mid- to high-end models, but launched its first small family car in China at the end of 2000. In the subsequent ten years, the Sail led the family car category, making up 80 percent of the Chinese car market with more than 20 million vehicles.

I saw the newly developed, not yet released Chevrolet Sail at the Shanghai Pan Asia R&D Center in autumn 2009. Equipped with

1.4-liter and 1.2-liter engines, the design was smart and the interior craftsmanship excellent. The body was the SII platform with exclusive property rights. It had Chevrolet's golden bow tie badge at the front.

Ding Lei, President of Shanghai GM, had a strategic vision that was always a step ahead of the others. He told me, "I don't think of us as an American company. We are a vital company based in China, but able to integrate the world's resources and constantly adapt to market needs. Our development team jokingly refers to themselves as the 57th minority group, where China has 56, and are called 'the never satisfied' with new goals always before them."

"Shanghai GM's medium-sized cars, the new Regal, the new LaCrosse, and the Cruz, are all very popular. Why do you want to take your chances with a small car with a narrow profit margin?", I asked Ding. "It's our basic strategy," he replied. "Small cars are most the widespread in China's auto market, and that's where there will be sustainable growth for Shanghai GM. Our raw materials and components are first-class internationally and can't be made cheaper. However, it's cheaper to buy 1 million sets of parts than 100,000. The Sail has that advantage over small-volume models."

I asked Ding, "Why not create another indigenous brand to accord with political trends since the Sail is indigenously developed?" He said, "You have to start with market realities when you do business. Of course, you can also call it an indigenous brand, but which is more welcome in the market, it or the century-old Chevrolet brand? The point of the new Sail is to create a good car that is in accordance with international development, achieves international quality, and is inexpensive. By putting the Chevrolet brand on it, more young people can enjoy the quality and service of an international big-name car and spend less."

Shanghai GM mastered indigenous development but borrowed international brands, which made Shanghai GM's market reach admirable. SAIC and GM decided to promote the Sail on the Indian car market.

Shanghai GM became China's most dynamic car company. One of the most convincing facts was that when General Motors was declared bankrupt by President Barack Obama in 2009, Shanghai GM wasn't affected by it. Sales grew nearly 80 percent that year. One important organizational change when GM reorganized was to move its international department to Shanghai. That showed respect and the expectations it had for Shanghai and Haipai style automobiles.

Shanghai Volkswagen R&D Center 1: The clay model showing the true shape of the car.

Shanghai Volkswagen R&D Center 2: Bench test simulating driving in complex road conditions.

Shanghai Volkswagen R&D Center 3: Acoustic test chamber.

Well prepared: SAIC Roewe starts with a bang

SAIC first acquired UK Rover as a base and then began building its own brand — Roewe in 2007, more than ten years after its joint venture started.

Roewe was able to stand on the shoulders of giants by relying on SAIC Group's foreign partners — Shanghai Volkswagen and Shanghai GM to nurture talent and accumulate funds. Sturdy SAIC walked tall as a flag bearer for high-end indigenous brands with outstanding research and development.

Authoritative neutral third-party data showed the Roewe to be surpassing some of the joint venture brands in popularity and reputation.

SAIC President Chen Zhixin said to me, "After dealing with foreign countries for so many years, we know that core technology of multinational companies won't be given to joint ventures. For example, you will never get integrated control of the whole vehicle, or control software or source codes for the engine management system (EMS), or transmission control unit (TCU), so we have to do it ourselves." SAIC invests about 5 billion yuan annually on its brands' R&D. Its Technology Center is headquartered at Anting in Shanghai and there is another development center in the UK.

To develop the Roewe brand, they first integrated the world's best resources, learned about overseas brands and technologies, and applied them to SAIC's own brand. Secondly, they kept introducing new and competitive products, armed with the latest product structure and technical content. Thirdly, all the parts and components they purchased were first-class, such as the Bosch EMS, Mario headlights, Johnson seats, and the same automatic braking system (ABS) as German models.

With their good technology and high value, monthly sales were in the tens of thousands for the 2009 Roewe 550, called the fully digital car, and the Roewe RX5, which became the internet celebrity model of 2015. Attention was drawn to the RX5, known as China's first internet car, for the use of zebra-like thinking to mix the traditional with the internet in its development. A zebra is a black and white striped horse and is the name of a joint venture between SAIC and internet giant Alibaba, which is a provider of internet car solutions. Two groups of people whose thinking and culture were completely different, came together, and integrated in a unique process. The Roewe became China's pioneer internet car.

When you test drive the RX5, one of the things you notice is the large central control screen and voice control system. The car is completely controlled by your mobile phone, and many functions such as air conditioning, sunroof, and music are controlled by voice and adapt to people's different accents through learning functions. At the same time, in terms of networking between cars and infrastructure, the RX5 is also interconnected with hundreds of parking lots in Shanghai. Parking spaces can be reserved in advance and paid for online via Alipay.

The Chinese car industry has gone from being one of the most worrisome industries in the fifteen-year WTO negotiations to one characterized by rapid development and global and market orientations. Joint ventures and indigenous brands are the two indispensable wings. Chinese people who were without cars for a hundred years can choose among almost all international and indigenous brands in the market today, at lower than international prices. It would be a blessing for the government and the nation if China's real estate and stock market could develop as healthily as the automobile industry.

The Shanghai hydrogen fuel cell car, which won third place in the 2011 Bibendum Challenge for New Energy Vehicles, was displayed in front of the Brandenburg Gate in Berlin.

Chapter 8

Marketing Automobiles: From Elementary School to MBA

Z hang Xiaoyu, who once served as the Director of the Automobile Department of the Ministry of Machinery Industry, recalled the period of the planned economy in the early 1980s, when a national car production plan was formulated each year for the following year. In Beijing, he only needed to go to three departments, the Central Organization Department of the CPC, the Personnel Department of the State Council, and the General Political Department of the People's Liberation Army, to know how many bureau-level cadres and how many division-level cadres would be promoted the following year. The standard for car allocation in those days was one for every four bureau level cadres, one for every two deputy ministers, one for each minister, and for military cadres according to mission. After calculating the total number, you only needed to fill in one form, make three carbon copies, and send it to the State Planning Commission for approval. Then the Commission would allocate resources. The amount of steel and rubber needed for each car and model was accurate to single digits. Zhang Xiaoyu remembered that only 3,400 vehicles were planned for in 1981.

1. How car sales evolved

Cars weren't for sale in those days

Chinese people had been building cars for more than a half century since FAW began producing Liberation trucks in 1953, but they had only been learning to sell them for 20 years.

In those days, Chinese cars were not manufactured to be sold, but were allocated according to a plan. As an important commodity, the plan for cars was set by the State Planning Commission, and production was arranged by the Ministry of Machinery Industry. Products were allocated by the Ministry of Materials and there was no private consumption whatsoever. It was all about following the economic plan.

This system continued until the beginning of Reform. When I began covering the Vehicle Bureau of the First Ministry of Machinery Industry in Muxidi in 1981, a section chief I knew was suddenly arrested. Newspapers reported that he had engaged in speculation and profiteering. In fact, it was for swapping some steel to help a company increase production of trucks outside the plan. Today it's like telling a tale from the Arabian Nights.

After the Reform period, the planned economy's dominance over the auto market gradually broke down, and the old sales network fell apart. Beginning in 1983, the government arranged for 10 percent of the plan to be sold by auto companies themselves, and expanded the proportion in subsequent years. In 1985, materiel departments established integrated automobile trading centers for sales within and outside the plan as a pilot for distribution reform in Beijing, Shanghai, Shenyang, Wuhan, Chongqing, and Xi'an. Sales companies also sprang up at various government levels.

Cars fully entered the market in the mid-1990s. The Automotive Industry Policy of 1994 emphasized the establishment of a manufacturer-based automobile sales system. Authorities decided to provide liquidity as a lever and focused on establishing agency distribution systems at FAW, Dongfeng, SAIC, Tianjin Automobile, and Nanjing Automobile.

The Beijing Administration for Industry and Commerce held a car show at the National Agricultural Exhibition Center every quarter. Organizations and individuals could purchase cars at the show with checks or cash. The sellers included materiel bureaus, auto repair factories, and individual entrepreneurs who acted as brokers. Private or individual dealers made up 80 percent of them. There were no storefronts at first, just a few cars sold in the empty spaces under Beijing's Second and Third Ring Road overpasses. This stage of selling was referred to as Beneath the Bridge, after a movie title of that time.

Cars began to be sold in stores in the late 1990s. Multiple stores and brands brought about the auto superstore. In the Asian Games Village and North Auto Markets in Beijing, the owners attracted traders by paying for the marketplace to be built, sub-letting vendor spaces, and providing services. Market enthusiasm was strong. When I went there, car fanatics swarmed around, gushing about models and prices. There were only a few sample cars in the marketplace and receipts were issued when cars were sold. Dealers took owners to a parking lot in the suburbs to drive the cars away. In 1998, sales at the Asian Games Village reached 50,000 vehicles with transactions of 5.7 billion yuan, the highest in the country.

The car market surged for ten years after China joined the WTO in 2001. Exclusive brand dealerships began to spring up following the international model. At the turn of the century, China's first 4S[1] brand store — Guangzhou Honda's specialized automotive sales and service store opened on Huangshi Road in Guangzhou. Audi, which mainly dealt in luxury models, followed closely with its 4S shops. They were designed to look like aircraft hangars — all metal and glass — and began appearing in every major city.

4S merged sales, spare parts supply, maintenance services, and information feedback into one store. Previously, car buyers had to deal with auto parts markets and roadside repair shops. Being cheated and getting inferior parts was a common occurrence. The 4S stores, which tied together sales and service, were favored by private car owners who gradually became their main customers.

The people who acted first reaped rich rewards. Zhongrunfa, Beijing's first Audi 4S dealership, began operations in 2001 with an

---

[1] Buying a car was a gamble until the arrival of the 4S dealerships. Short for Sales, Service, Spare parts and Surveys (customer feedback), 4S stores authorized by foreign and domestic carmakers arrived in the early 2000s and quickly began to set new standards in the car-buying market.

investment of 30 million yuan. It recovered its costs that same year. In 2003, the Guangzhou Honda factory authorized its dealerships to increase the price of the Honda Accord, which was in high demand at the time, by 30,000 yuan. The profit forecast for BMW meant that there were more than 3,000 competitors for the first group of 24 dealerships. The shenanigans they played among themselves could only be imagined.

FAW-Volkswagen's first sales team

The sales system which automobile manufacturers established, and the pattern by which factories controlled sales rights began at FAW-Volkswagen in the mid-1990s.

For many years, cars and trucks produced by FAW Group had been sold together. By the mid-1990s, this crude mode could no longer adapt to market competition. Annual sales of the Jetta plus the Audi hovered at just over 20,000 vehicles. With a production capacity of 150,000, FAW-Volkswagen suffered serious losses because they couldn't sell their products.

To get out of the predicament, FAW decided to spin off its car sales. FAW-Volkswagen Sales Company was established in April 1997. Zhou Yongjiang was appointed General Manager. The key members of this sales company later became sales bosses of the various brands in the Chinese car industry. The first people who entered the sales company couldn't even formally transfer into FAW-Volkswagen. At the beginning, they got 500 to 1,000 yuan per month in living expenses.

The new company was overwhelmed by the task of increasing its sales. FAW Group had set a difficult goal, suddenly jumping to 40,000 from the 26,000 vehicles of the previous year. After a year of hard work, the company increased sales by 57 percent to 41,000 vehicles, an impressive achievement. Moreover, relying on the pull of its products, FAW Volkswagen turned itself around that year, achieving profit of 195 million yuan for the first time.

I first heard of integrated marketing from Zhou Yongjiang in 1998. This integration was an enhancement of the weakest link. It was led by different levels of customer demand, and marketing was a process of creating value and establishing a brand image.

Marketing has no fixed style. It is an art and there is a lot of room for creativity. When the sales company started up, the only product it had was the Jetta. It came in three colors — red, white and blue, with manual transmission only. They worked hard to make Jetta into a brand. Its engine, appearance, and interior were entirely reworked in five short years. Jetta was the first to introduce metallic paint, the first to be configured with ABS, the first with an automatic transmission, the first to have 20 valve EFI engines, and the first to transform its appearance. These features were used to stimulate market interest. Models such as the Urban Sunshine and the Urban Spring were derived from the Jetta. FAW-Volkswagen also made a grand spectacle of a taxi driver in Zhuhai who went 600,000 kilometers without a major overhaul, by giving him a car and an award. Using common sense to integrate the demands of market competition, the Jetta was gradually acknowledged by consumers as sturdy, durable, and affordable.

The sales company was also a learning organization. Any good ideas and experiences from outside were brought together as shared knowledge. An executive told me a story: "If a Chinese and an American came together and exchanged a dollar, each person still took away a dollar. But if the sales experts of the two countries got together and exchanged experience, when they went their own ways, each person took away twice as much."

German consultants had been brought into every department of the sales company since 1998, later holding positions corresponding to the Chinese department heads and managers. The selling point of the Bora, a driver's car, was the idea of VW's consultant Smith in 2001. It emphasized its superior power and control, and avoided the shortcomings of the rear seat.

How did Shanghai GM stay productive?

Shanghai GM constantly engaged in innovation in China. Industry praised it for how well it marketed and increased sales of American cars that had no outstanding selling points, and it made the most sales for many years.

Sun Xiaodong and Liu Yuehai were marketing partners from China and the US. In ten years, they went together from sales director to company vice presidents. From the beginning, GM unreservedly brought nearly a hundred years of successful marketing experience to China. Shanghai GM's marketing system achieved two industry breakthroughs. One was to transform service into a brand, and the other was to standardize dealer operations processes.

Chinese people had no concept of service when they began selling cars. GM's Saturn Company brought branded service to reality early in the 21st century, putting emphasis on the moment when customers came into contact with products and began to personally experience brands. This customer focused concept had just been introduced in the global automotive industry. Sun Xiaodong attended training at the Saturn Corporation in Nashville, Tennessee. Afterwards, Shanghai GM launched the Buick Care service brand. It offered transparent charges, booking services, and 24-hour service, personalizing the sales and repair experience. Shanghai GM also overturned the old concept where manufacturers dominated dealers who were contractors. It promoted dealer operations standards (DOS) and regarded dealers as close partners. This created the best profit environment for dealers by using advanced standardized management models and financial support.

When Buick's New Century came to China, the Chinese had already been brainwashed by the exquisite interiors of Japanese cars. The comfortable, wide, and quotidian features of American cars weren't appreciated by Chinese consumers. The Regal, a local development project to replace Buick's New Century model, had just started, and interior design at the Shanghai Pan Asia R&D Center was affected by the aesthetic demands of Chinese consumers.

When Shanghai GM launched the Buick Regal in 2002, it kept improving the quality of its materials, its color schemes, and its workmanship. Moreover, it underwent thoroughgoing changes as it sought to distinguish itself from the New Century. GM Deputy Chairman

Bob Lutz was pushing to make GM's interiors as refined as those of Japanese cars. When he saw Shanghai GM's new Regal, he was over-joyed and immediately had it brought to Detroit. It was placed on exhibit in the entrance hall at GM headquarters, and managers were told to go and look at it.

Shanghai GM's marketing director got together with the engineer-ing department at the Pan Asia R&D Center twice a week. Translating market demand into engineering language directly affected the devel-opment and improvement of new cars.

The subprime mortgage crisis caught up with GM in the US in the spring of 2009. The Shanghai GM marketing team still got Buick's New Century replacement off to a good start and further improved its multi-brand composition.

There had been many successful examples of dual-brand or multi-brand attempts in China's auto market in recent years, such as Dongfeng Peugeot-Citroën Automobile Company's Citroen and Peugeot, Shanghai Volkswagen's Volkswagen and Skoda. But Shanghai GM's Buick, Chevrolet, and Cadillac multi-brand strategy was the most successful.

By the beginning of the 21st century, Buick was an old-fashioned, subsidiary brand of GM. The average age of Buick owners was 68 in the North American market, and sales had shriveled. It was Shanghai GM that resuscitated the Buick brand in China, deliberately creating a medium- and high-class classic brand that was safe, quiet and had a graceful, flowing design. It was no exaggeration to say that without Shanghai GM's successful reshaping of the brand and selling more than 90 percent of them in China, Buick, like Pontiac, might have been discontinued.

I attended the GM Global Media Conference in Nice in September 2004. On the last day, Wagner announced GM's global brand strategy, changing many of GM's brands into regional ones. He announced that he would turn Chevrolet into an international brand. Ten new Chevrolet cars launched in the previous 20 months were displayed on a golf course that evening. They showcased the developmental strength gained through the Daewoo acquisition. There was no shortage of smaller models among them, including the Matisse which became successful in Europe.

Selling a new car every 7.2 seconds was the record set in the US during Chevrolet's heyday. Baseball, hot dogs, apple pie, and Chevrolet were symbols of an era in American life. The Buick is like a formal suit and tie, while the Chevrolet is like casual attire, more sporty and younger. That was the biggest difference between the two brands.

Cadillac became a high-end brand in Shanghai GM's multi-brand strategy.

Dongfeng Nissan: Refining salt from seawater

Dongfeng Nissan suffered by being an outsider when it came to producing sedans under the Dongfeng Group.

However, it proved to be a dark horse in the Chinese auto industry. Starting out with the production of the Fengshen Bluebird, the product line ranged from the A0-class Lavina and March, to A-class TIIDA and Sylphy, Qashqai and X-Trail SUVs, and the new high-end B-class Altima. Its ten plus models all sold well. It stood out among Japanese manufacturers with sales reaching 1.16 million units in 2018.

The Dongfeng Group turned its attention to Guangdong, establishing an entity under its Southern Business Division in the 1990s to produce components and engines and do commerce. The company used its first 23 million yuan as seed capital to build its Fengshen brand in 1999. In cooperation with Taiwan's Yulon Motor Company, it used its working capital to establish a factory to assemble smuggled cars based on the first generation of Nissan's discontinued Bluebird model. It jointly developed a third-generation Bluebird with Yulon. Supply couldn't meet demand and it made profits of 4.5 billion yuan in three years, getting the Dongfeng Group out of difficulty.

Guangzhou Fengshen Automobile Co., Ltd. was established in December 2001. At a time when automobile manufacturers could easily invest tens of billions of dollars, Fengshen embarked on a new path with minimal investment and quick results.

The success of the Fengshen Bluebird allowed Nissan, which was originally sitting on the sidelines, to see the potential of Fengshen and it proposed a joint venture. At this critical juncture, when the State's general rule was limited support for limited cooperation and major support for major cooperation, Dongfeng Group and Nissan formed a comprehensive joint venture in June 2003. Fengshen also became the Dongfeng Nissan Passenger Vehicle Company.

It took Dongfeng Nissan a year and a half to build a new plant with an annual output of 180,000 vehicles in Huadu, Guangzhou. The Xiangfan base for the production of high-end cars such as the Altima, the Zhengzhou base for the production of SUVs, and a second factory at Huadu were then built. The TIIDA, the Converse, the Qashqai, and the new Altima became solid gold for Dongfeng Nissan with monthly sales of over 10,000.

Dongfeng Nissan's sales remained as robust as they were at its founding. Its dedication could be seen in the brand experience of its marketing.

Dongfeng Nissan participates in national auto shows in Beijing, Shanghai, and Guangzhou every year. And it doesn't pass up local auto shows like Fushun in Liaoning and Anyang in Henan. It attends more than 70 first class local auto shows a year. The company also has a year-round product display caravan team for beer, clothing, and kite festivals. Its products are displayed in the middle of a stage when the six caravans form a triangle.

However, displaying products is not the sole purpose of auto shows and the caravan team. The staff busily collects on-site evaluations of visitors, screens prospective customers, and calculates order rates. In addition, netizens who click on Dongfeng Nissan's website for individual model information and learners who are at driving schools are all potential customers of Dongfeng Nissan. The company's call center has seven or eight hundred positions and is always available to answer inquiries from customers. The data is fed back into the company's database and the program selects customers who have a high intention to purchase. Sales experts contact them in a targeted manner, in line with their purchasing behavior.

In the words of Dongfeng Nissan, this process is like extracting brine from seawater and then refining it into salt. Many of the 500,000 to

The sleek and beautiful Dongfeng Nissan March.

600,000 sedans which Dongfeng Nissan sold each year may have found their owners in this way.

2. Transforming China's sales system

Brand Management Measures in a tangle

It is difficult to offer a simple evaluation of the Implementation Measures for the Management of Automobile Brand Sales (hereafter referred to as Measures), which were put into effect from April 2005. They had been widely denounced, but still thoroughly implemented.

The Measures drew upon the prevailing model of international car sales, covering all brands of cars sold in China. One part of the system was the manufacturer's own sales company. The other was a specialist dealer, authorized after undergoing assessment, commonly known as a 4S shop. Even though they were outnumbered, manufacturers still maintained a strong position, controlling the standards and operations of the specialist dealers.

According to the Measures, unauthorized companies could not sell cars. This move changed the chaotic and scattered nature of car dealerships, protecting consumers' rights and interests, and enhancing service awareness of car dealers.

When car sales were in their infancy, a car owner often had to deal with auto parts centers and roadside repair shops after purchasing a car. The new mode guaranteed the consistency of the manufacturer's trademark, sales, service, maintenance, and genuine parts supply from dealerships, thus achieving the long-term goal of building the brand. To maintain this network, manufacturers had to invest a lot of money and manpower to fund and train dealers.

4S stores integrated the functions of sales, spare parts supply, maintenance services, and information feedback for more than a decade. Chinese private car owners had accepted this combination of sales and services.

Most financially stable dealers were content to begin with. They scrambled to become dealerships of famous brands, spending serious money to buy land to build stores. The cost of construction for a 4S shop could go as high as tens of millions of dollars. Some dealers' profit margins began to shrink in the increasingly fierce competition, and they ripped off consumers to make up for these losses. Even worse, manufacturers forced stock on dealers to reach annual sales rankings, tying up a lot of the dealers' capital. The 4S sales model began to come under fire for monopolizing the market among other things.

The Ministry of Commerce implemented its new Automobile Sales Management Measures on 1 July 2017. It removed the term "brand" from the name. The key point being that dealers didn't have to be authorized by a manufacturer in the future.

A dealer's daily necessities

Beijing Yuntong Group is a large-scale automobile sales group. It has seven 4S brand stores in the Beijing Economic and Technological Development Zone, which is home to the world's top 500 companies. They are mostly for mid-to-high-end brands such as Audi, BMW, Jaguar Land Rover, Infiniti, Buick, FAW Volkswagen, FAW Toyota, Dongfeng Honda, and Skoda.

Yuntong was a repair specialist which became a brand dealer after the big surge in 2002. Its boss was a man with a powerful strategic vision. At that time, there were private auto repair factories in Beijing like Sanlian and Tengfei. They did well, but when the business model changed, they didn't change with it.

Things were easy for the first 4S stores like Yuntong, as long as they didn't make any mistakes. The Fifth Ring Road hadn't been built in 2002 and land around Beijing was still very cheap. But these days automobile companies stipulate at least 10,000 square meters of floor space for their stores. It takes two or three hundred million yuan just to buy the land. Renting land is even riskier. Building a reputation can take eight to ten years. When the lease is up and the landlord takes back the land, you don't have anything to show for it.

Operating a 4S store isn't just selling cars. Profit margins for selling cars are very thin. Money is made by maintaining good customer relationships, and relying on routine maintenance, repair, insurance, and other after-sales services.

Factories give a commission of six points for sales of cars valued at 100,000 yuan. However, in order to sell more cars, the 4S shop gives consumers points for most of them. Points are also deducted for factory assessments, so each brand's average annual profit is only one or two points. Compare this with the three or four points in the boom year of 2009, which earned FAW-Volkswagen twenty to thirty thousand yuan per vehicle. But before that, there was also a time when each Jetta sold for a loss of 7,000 yuan.

There are about 10,000 4S auto stores now. In the past two years roughly one-third suffered losses, one-third barely survived, and one-third were profitable. A 4S shop for a low-end brand needs to sell around 1,500 vehicles a year. The mid-range brand stores have a profit and loss point of about 1,000 vehicles, while high-end brand stores are at 500–600 cars. But even in a city like Beijing with a developed auto market, only about half of the dealers can achieve that.

In the first year that a 4S store opens, it has to rely on sales and won't make money. By the third year, sales and services profits each account for half. And it becomes self-sustaining by the fifth year when services account for 78 percent of fixed costs. Nowadays, Yuntong's several stores in Beijing are generally mainstream joint venture brands, with single store sales ranging from 2,000 to 3,000 vehicles annually. Business is booming at Yuntong's first Beijing store for the imported Infiniti brand. With annual sales of 800 vehicles, its performance is first class.

4S shops are happiest when crowded with customers. Yuntong has its greatest concentration of 4S shops in Beijing in the Yizhuang Economic and Technological Development Zone, home to the world's top 500 enterprises and filled with high-end residential areas. Manufacturers aren't considered first-line brands if they don't have 4S stores there.

3. The value chain extends beneath the wheels

When will we sell more used cars than new ones?

It is often seen as a sign of maturity in a nation's auto market when more used cars are sold than new ones.

When driving in Japan, you will often see flags fluttering by the roadside with the words "USED CARS" in big letters. Neatly arranged among the array of flags are cars of all colors, washed as clean as new. These are Japan's exclusive agencies for the sale of used cars. There isn't any chaos and noise like in Chinese used car markets. Everything is clean, quiet, and calm.

The sale of used cars in the US and in Europe is similar for the most part to that in Japan. But second-hand cars for sale in Japan are quite beautiful, and they are 70–80 percent new. The US has both older and newer used cars. A few hundred dollars might buy a 20 years old car. That was the experience of many Chinese students there when they bought their first car.

When annual sales curves of new and used cars in Japan are compared, the curve for new cars is a mountain. The peak was in 1990 when total sales volume was 7.78 million. A downward trend followed, falling to 5.78 million in 2002. In 2009, it fell further to 4.29 million. The total sales of used cars is an upward slash. After the 1990s, sales went in the opposite direction from new cars, and there were several times as many transactions.

The same is true in the United States. Total sales of new cars was 16.7 million in 2003, while the sales of used cars reached 43 million, more than double the number of new cars.

That is because car consumption is close to saturation in Japan, the US, and developed countries in Europe. Almost every family has a car, and not just one car. The main function of cars is as a means of transportation. Additionally, with good roads and standardized maintenance, conditions for used cars are similar to those of new cars. Chasing new models, or buying an affordable used car, is based solely on personal economic conditions and preferences. Especially in an era when the economy continues to be sluggish, good quality, low priced, second-hand cars are naturally popular.

The used car market in China is changing. The first used cars in the open-air markets were filled with dust. Other than the year of production and total mileage (inevitably set dishonestly), there was no way of knowing if the machine was faulty, or if it had been in a major collision, or anything else. Buying a used car was a matter of luck. Car buyers, especially private ones, often held back the information that didn't match up.

There are two-tier markets for sales of used cars in other countries. In Japan, I visited the Hanaten Auto Auction (HAA), a used car auction market in the Tokyo suburbs. HAA is a large network throughout Japan and has a membership system. The primary market acquisition is carried out by HAA. Professionals inspect and assess the old vehicles, repair existing faults, and have a professional studio take photos of the car, before putting them online for auction. Auctions are limited to members, and members are mostly retailers of used cars. The auction site I visited had 300 seats, and the long table in front of each seat had a numeric keypad. There were two large screens at the front of the auction house, which quickly scrolled through the photos and related information about each car being auctioned. Participants submitted bids electronically and the highest bidder won. There were no auctioneer's calls or banging of the gavel, but people on the sidelines were spellbound. A single auction house could conduct 100,000 trades a year!

Outside China, dealers of some brands buy used cars by exchanging them for a new car of the same brand. From the dealer's point of view, they keep customers who buy one new car after another. From the customers' point of view, they only have to make up the difference in cost from their own second-hand car in the replacement process, eliminating the bother of having to buy and sell a car separately. Moreover, the seller gets a better price for his second-hand car and it can be resold for a good price.

Some forward-looking, mature Chinese car brands have also seen the value of this. Buick, Audi, and Toyota were the first to launch trustworthy used car businesses around 2004. After comprehensively inspecting and repairing second-hand cars received as trade ins, they are placed in branded used car stores at reasonable prices. As opposed to ordinary domestic used cars, these not only have the quality approval of the brand, but also a warranty for a period of time or number of kilometers. This gives assurance to used car buyers.

The sales ratio of used to new cars in China is only 1:3. This gives scope for development compared with the ratio of 3.5:1 in developed countries. Establishing a modern, standardized, and reliable second-hand car market will certainly benefit manufacturers and consumers. Perhaps after three to five years, a system for both new and used cars will emerge in China.

A used car store on a street corner in Japan. Second-hand cars sell well.

Geely Taobao: Online shopping is not just a dream

I first heard about buying cars online when I was in the US in 2000. Although shopping online for fast-moving consumer goods has become commonplace, online shopping for high-value, durable consumer products that emphasize a hands-on experience, are in use over a long period, and need years of maintenance, like cars, have grown slowly. But there have been many attempts in Europe and the US. After GM's recovery, one of the first things it did was to work with eBay to create an e-commerce model for online car purchases.

In China, with the world's largest number of internet users, a trial of online car shopping began in 2010.

Taobao conducted a Mercedes-Benz group sale on 9 September 2010. The small Mercedes-Benz Smart Car with an original price of 176,000 yuan was bought by the group at a discount of about 40,000 yuan. The event was immediately popular, and the first Smart car was sold in 24 seconds. Fifty-five of the cars were sold in six minutes, and all 205 were sold in three and a half hours.

Even if it was just a one-time promotion, it still demonstrated the huge potential for an immature car sales format.

Liu Jinliang, Geely Group Vice President and General Manager of Sales, was among the first to join Li Shufu's team. He became an active promoter of online sales of domestic cars. Geely opened its flagship store on the Taobao On-line Mall at the end of 2010, becoming its first car sales company. It was the first legal and lasting "marriage" of Chinese car companies with the online shopping giant.

The link between Geely and Taobao could be because they were both headquartered in Hangzhou, or perhaps because both Li Shufu and Ma Yun (Jack Ma) were always trying something new. Liu Jinliang said, "We are selling cars, not toys. At the moment, you can't hear a knock at your door, and go out to see that the new car you purchased online has been delivered by express delivery." Buying a car involves test drives, licensing, inspections, insurance, and maintenance servicing. It's hard to do that over the internet, so physical stores have to be part of it.

Online shopping has gained a foothold. Firstly, 80 percent of people today get information on cars through the internet. They can find out more about a particular model on the internet than in the store. More details can be displayed through 3D and cutaway diagrams. Secondly, as the Internet Real-name System is gradually implemented, dealers can determine what information their customers need and accurately communicate with them. Thirdly, it makes it easy to conduct transactions in different locations. For example, a successful young person can purchase a car for his parents in their hometown far away. He can send the new car with customized services to his parents through online shopping.

Shanghai Volkswagen has also launched its Skoda e-Purchasing Center. With real-time 3D tools, users can customize their purchases and make payments online. When the order is placed, the Skoda dealer will have a sales consultant provide a full range of services. More and more Chinese car companies have expressed interest in selling their cars online.

Wheels churn out new wealth

The continued consumption by a car throughout its lifetime is the biggest difference from general fast-moving consumer goods. It far exceeds the purchase price and is the biggest source of revenue in the automotive industry chain.

The industry chain extends beyond just repair and maintenance. Insurance, financing, advertising, detailing, car audio, tourism, and even intelligent transportation will become large, independent industries in China.

GM China once gave me some statistics. In the United States in 1932 and Brazil in 1986 (when their respective total car production was close to China's in 2000), an automobile manufacturing plant provided 11 jobs in upstream and downstream industries for each position in its factory. One out of every six people employed in the US serves the automobile and related industries.

Take car maintenance as an example. Car owners outside China concentrate heavily on maintenance and have fewer repairs. It's like focusing on good health and less on going to the doctor. Thus, the car maintenance market has increased significantly. The annual turnover of car maintenance in the US is more than US$100 billion, accounting for 80 percent of the car warranty industry, reducing the write off rate to 21.7 percent.

Auto insurance accounts for 60 percent of the world's non-life insurance. When Japan was growing a car culture in the 1960s and 1970s, car ownership increased four-fold, while income from premiums multiplied by 11. Even in China, auto insurance ranks first in property insurance.

In developed countries, consumers buy cars through credit and leasing. This is the case for the bulk of car sales. The proportion is 92 percent in the US, 80 percent in the UK, 75 percent in Germany, and 44 percent in Japan. As well as banks, all major global automobile groups own their own finance companies. Automobile financing has become a major source of profits, more profitable than manufacturing.

Around the time China joined the WTO, automobile loans became a minefield. A number of commercial banks initially tested the waters. But because car buyers' credit couldn't be checked, they incurred a lot of bad debts, and beat a hasty retreat when the financial system was consolidated. After several years of waiting, the first automotive industry finance company finally obtained a permit. SAIC-General Finance Company was approved by the China Banking Regulatory Commission (CBRC) and started operations on 4 August 2004. It began in Beijing and Shanghai with registered capital of 500 million yuan. Its first step was to provide loan services for the Buick brand. Then it gradually expanded to other brands in the group and to other car companies.

Following closely on the heels of Shanghai GM, Ford China, Volkswagen China, and Toyota China also received approval from the CBRC to provide financing. This brought strong, experienced competitors into the alluring auto finance market.

Auto finance companies can provide support to car dealers. When sales are sluggish, inventories build up, and cash flow decreases, major banks are often reluctant to approve loans. Auto finance service companies can give more support than a bank whose fortune is tied to the dealer.

There is some way to go before Chinese people have a car culture, and some of the extended services around cars are still unfamiliar. Driving on the highways in the US, there are probably more motels than hotels. Car owners can enjoy convenient services while sitting in their cars, such as drive-in fast food restaurants, and drive-in cinemas. Professional audio and entertainment systems in your car may be more expensive than the car itself. Car advertising which flashes by the window may constitute the world's largest advertiser. Satellite positioning systems in cars and electronic toll collection systems are not merely new crazes in the digital age, they are a small part of intelligent transportation.

4. It takes more than a day to create a brand
Cars also have emotional appeal

It's clear that creating a homegrown car is easier than developing a brand.

In the early 1980s, a billboard which got your attention as you drove from the newly built Capital Airport into Beijing declared that "When you get to the foot of the mountain, the road appears. Where there is a road, there is a Toyota." The slogan came from a Japanese student at Peking University and soon became known everywhere. The Toyota Crown was the epitome of Japanese cars. The classic, conservative look which still appeals to older Chinese people comes from the large numbers of Toyotas imported in those days.

According to the Toyota China Chief Representative, the price for each Toyota Crown imported in 1973 was 5,700 yuan, settled in US dollars. It seems cheap today, but was equivalent to ten years salary of a skilled mechanic. The price of the Crown coming off the line in Tianjin in April 2005 was between 320,000 and 480,000 yuan. The technology and luxury of the old and new Crowns can't be compared.

Automobile products are characterized by the history, culture, and national characteristics of the countries which produce them. German cars go all out in the pursuit of new technologies, continuous improvement of intrinsic quality, and excellence in power and handling. Japanese cars have balanced, round shape and meticulous interiors. They use mature European technology and cost reduction is a priority. American cars are luxurious, stylish, comfortable, and relaxed, but distinctive. French cars are bohemian, elegant, eclectic, and dare to try a variety of technologies and configurations. British cars are steady, old-fashioned, untrendy, and self-satisfied.

I have visited Toyota many times and have tried to gain an in depth understanding of Japanese vehicles. I have a detailed understanding of the Toyota production methods which are regarded globally as models for quality control. And I have seen factories that continue to use equipment from the 1930s to make parts which are up to present day standards. I have also visited the Toyota Technology Museum and Odaiba Science Park that connects with the younger generation. However, what made me understand the cultural differences between Japan and China was bathing in an onsen (Japanese hot spring bath).

I have bathed in hot springs in Japan many times. There are traditional baths, as well as luxurious, modern ones. For the Chinese, bathing is like washing a muddied radish in water, leaving the dirt from the body in the water, while the Japanese first shower and wash themselves thoroughly before going into the hot springs. Attention is paid to the surroundings of the hot springs, and some are even set in traditional gardens. I went to a hot spring pool set on the rooftop of a mountain hotel, surrounded by pine forests. The sun rose and wind whistled through the pines, blowing snow from the needles onto my face as I soaked in the hot water. I realized that hot springs don't wash the dirt, but the soul. Japanese cars blend into this culture, which is why the quality and beauty of Japanese cars are what they are.

German cars pursue technical perfection. German Volkswagen has a slogan — "For the love of the automobile." Its luxury brand Bugatti, at a price of 2 million euros, has an engine with up to 1,100 horsepower. From a commercial point of view, so much power is difficult to fully exploit on the road, but its original intention was to challenge the limits of automotive technology. This is Volkswagen. Many automakers are making cars that are a practical means of transportation, but Volkswagen is building precision technology. Their engineers' thinking is to make perfect cars, and only then consider controlling costs.

Maybe this has something to do with the German character. Dr. Zhang Xinxin, Vice President of Volkswagen China, gave me an interesting example. When FAW-Volkswagen built its production line, an electrician was installing a distribution box, and a German technician came over to check the wiring. The Chinese person was surprised when he took out a level and checked if the wires were horizontal and vertical! This is the nature of the Germans, perhaps related to the education and training they have received since childhood — the pursuit of perfection, without any loose threads.

For hundreds of millions of car owners around the world, cars are more than just machines made of steel, rubber, and glass. They have a certain spiritual quality, and just like pets become beloved. Some people like power, some like beautiful interiors, some like rough edges and corners, and some like smooth shapes. This diversity of desire gives car manufacturers unlimited opportunities.

I participated in GM's Product Symposium, in Santa Barbara, California, in 2002. One seminar was "What the Fashion, Entertainment, and Automotive Industries Have In Common." There were three guests on the stage, a representative of the Italian jeweler Bulgari, a prominent Hollywood film director, and General Motors Vice Chairman Bob Lutz, known as the Product Czar.

Lutz said in his opening remarks, "Today's car is not only an extension of peoples' bodies, but also an emotional demand beyond rationality. In this way, a watch is not only a tool for calculating time, but also a crystallization of elegant fashion and fine work."

Bulgari has jewelry stores are all over the world, and says that its job is selling dreams. "Classic creation takes time. The Cadillac, for example, has been built for a hundred years. Good design stimulates the impulse to buy. The impulse is often not rational, but emotional," said the Bulgari representative.

The film director said, "The car industry is talking about the role of emotions. The film industry should pay attention to that. Emotions can be far-reaching, and passion comes first in good car design. Focusing on emotions in car design is something to take delight in."

There is an old saying in China — "To raise its social status, a family must dine better for three generations, and must dress better for four." A brand is like a family's social standing, it is forged over time. Building a car brand takes decades and a lot of investment. It has to be supported by the development of a complete model series. Mercedes-Benz has an unshakable position among global car brands. With its technology and individuality, it has been at the high-end of the car industry for more than 130 years since its birth in 1886.

Among global brands, Ford, Buick, Chevrolet, Audi, Citroen, Fiat, Skoda, Toyota, Volvo and Jaguar are all proud of their uninterrupted history. They have all existed for 70 or 80 years, and some even over 100 years.

However, brands are not static. They adapt to changes in markets and over time. To emphasize its humanity, BMW launched a new brand strategy in China — "Joy is BMW." It abandoned its free-wheeling, money worshiping image to make itself more attractive.

The history of China's indigenous car brands is still short. Apart from the Red Flag, most are only ten or so years old. Their brand premium is without a doubt lower than those 100 years old international brands. That's nothing to be discouraged about. Korean car brands such as Hyundai Kia Group have successfully entered the mature markets of the US and Europe and have grown independently for half a century. While it is ranked among the top ten auto companies in the world, with product quality repeatedly ranked highly by JDPOWER in the US, their product premiums are about 15 percent lower than Volkswagen, Toyota, Ford and others. Daewoo, which was acquired by GM and placed under Chevrolet, hasn't been able to sell at high prices because of its Korean heritage. In fact, Japanese goods had a poor reputation in China in the first half of the 20th century. Japan held its first industrial exhibition in Beijing in 1961. Only three-wheeled vans were exhibited. After more than half a century of hard work, Japanese autos, especially with their improved product quality, gradually changed that image.

Announcing brands with fashion shows

Cars have become a global industry, with one out of every three cars being exported. More money and talent are used in announcing and marketing new cars than for any other manufactured product. World-famous tourist destinations are chosen for new car launches and promotions, giving national automotive media striking backdrops for their stories and photos. This is especially true for luxury brands.

Mercedes-Benz launched its first M-class SUV in North America in 1997. The site selected was the US Space & Rocket Center in Alabama. The car was displayed between a lunar module and a satellite. At night, the words "BENZ M CLASS" were marked out in lights on a huge space shuttle fuselage. It was a striking sight.

Dubai has the world's most pleasant beaches, and the world's tallest and most luxurious seven-star hotel. A vice president of BMW's sales department said, "The BMW 760 is the flagship of luxury cars, and first-class cars need a first-class backdrop." I went to the launch of the BMW 760 in Dubai in 2002. It included visits to gold and diamond workshops, a cruise on a luxury yacht, camping in the desert, and visits to the Royal Racecourse and Palm Island, a group of super luxury villas built on the sea. Wealth, speed, elegance, and range all reflected what the BMW stood for.

It was no exaggeration to say that the launch of the small Fiat New 500 in 2007 was at the level of the opening ceremony for the Olympic Games! The organizer for the commemoration of the 50th anniversary of the car was the same company that had planned the opening ceremony of the 2006 Winter Olympics in Turin.

The River Po, the largest river in Italy, passes through the city of Turin. The main venue for the party was two or three kilometers along the river. Motorboats cruised past the VIP stands with costumed performers. The night sky was filled with grand fireworks displaying the numerals 500. The water stage was wonderful, especially a large group of men and women wearing silver tights climbing a three-story metal frame with their bodies linked together to form a huge model 500 which was traced by spotlights as they hung from the tower. It attracted half of the eyes in Turin. It was Italian artistic inspiration and was overwhelming in its expression of the car brand.

The Beijing Olympic Games, on which the strength of the whole country was exerted, were held a year later. I didn't approve of Zhang Yimou's design for the opening ceremony. In the large space where 100,000 people were watching, there were often some ant-sized actors performing individually. People in the audience couldn't see them clearly. Which had more impact: a Peking Opera performance by four marionettes or a car model magnified a thousand times suspended in the air?

At the end of the Second World War, a group of cars developed for the civilian population became classics. They were the German Volkswagen Beetle, the French Citroen 2CV, the British MINI, and in Italy the Fiat New 500. When older generation Italians married and had children, the 500 went with them like a family member. On the day of the launch event, thousands of old 500's paraded through Turin. People were as mad about them as they were about football. One hundred thousand city residents took part in the grand event that went on throughout the night.

Long, boring speeches are unnecessary at a successful new car launch. On that day, Fiat CEO Sergio Marchionne held hands with a group of children on the stage. They talked about what cars should be like in the future. The children's innocent insights, such as cars being used to chase girls, caused a lot of laughter. Marchionne said, "Five million people participated interactively in the design of the Fiat New 500. We're a hundred years old, but we still look at tomorrow's cars with a child's eyes."

A brand is overpowering in its cultural penetration.

With the entry of mainstream global automakers when China joined the WTO, new cars, both locally manufactured and wholly imported, came out more frequently. There was a gradual integration with international markets.

The market entry of the new Shanghai GM Buick Regal in 2002 was a turning point. A symphony orchestra played magnificent music, and then the curtain was drawn on a deep corridor in the newly opened Shanghai Convention and Exhibition Center. Headlights came on in the darkness as three Buick Regals powerfully and majestically approached.

Later, new car launches of domestic brands became creative makeovers of luxury shows. I saw more than a dozen dances for the Audi A6L set off by fireworks. I saw the Nanjing Fiat Siena leap out of the misty waters of West Lake. I saw the Mercedes-Benz S-Class on the same stage as Zhang Yimou's Turandot in Beijing's majestic Taimiao[2]. Another innovative effort was the participation of the Chery X5 in the

---

[2] The Imperial Ancestral Temple, or Taimiao of Beijing, is a historic site in the Imperial City, just outside the Forbidden City, where during both the Ming and Qing

Paris Dakar Rally. The manufacturer arranged for some domestic media to fly to Argentina to boost morale[3]. It seemed that Chinese were as clever as foreigners.

Test drives everywhere

In addition to exciting displays, major international automakers emphasize test drives by professional journalists from all over the world before the launch of a new car. They let them try out new cars and convey their experiences to consumers in words and pictures. These journalists are car experts, and their comments directly affect consumer choices around the world.

Test routes arranged for journalists are more complex and challenging than the road conditions encountered by ordinary consumers. In 2002, when General Motors arranged a test run for the Hummer H2 large off-road vehicle in Santa Barbara, California, it rented a hill and invested hundreds of thousands of dollars to simulate rugged road conditions like steep slopes of forty or fifty degrees, mud, hairpin turns, scattered boulders, fallen trees, and rough roads. Only this type of off-road vehicle could face such conditions. It was the best display of the Hummer's robust power and unique control. Although you could rely on the car's performance, it wasn't without risk. When I saw the ambulance and fire truck hidden in the shade of the mountain, I realized there was danger along with the fun.

I test drove a Saab 9-3 under the glimmering northern lights in the ice and snow of the arctic circle in Sweden. I experienced the spin and drift when the Electronic Stability Program (ESP) went out of control. I also drove a Mercedes Benz SLR luxury sports car at 270 km/hour

Dynasties, sacrificial ceremonies were held on the most important festival occasions in honor of the imperial family's ancestors.

[3] The Dakar Rally (or simply "The Dakar"; formerly known as the "Paris–Dakar Rally") is an annual rally raid organized by the Amaury Sport Organization. Most events since the inception in 1978 were staged from Paris, France, to Dakar, Senegal, but due to security threats in Mauritania, which led to the cancellation of the 2008 rally, events from 2009 to 2019 were held in South America. The 2020 edition is being held in Saudi Arabia.

from Cape Town, South Africa to the Cape of Good Hope. When I stopped at a rest area and opened the carbon fiber gull-wing door, it attracted the other motorists like a magnet. In Scotland, with its richly colored hills and castles under bleak, overcast skies, I drove the Land Rover Discoverer 3 into a peat bog at the seaside less than knee deep, and even flew along the shoals to stir up waves in the sea.

Chinese auto companies and the media quickly surpassed multinational companies in promotion by test drives. On National Day in 2009, we drove the sixth generation of the Golf produced by FAW-Volkswagen across the Alps, on the rugged mountain roads between Switzerland and France. This was a part of FAW-Volkswagen's trans-European Golf Perfect Experience Tour. The new Golf could smoothly and powerfully maintain speeds of 120 kilometers per hour when climbing and negotiating curves. Its fuel consumption was only about 5–7 liters per hundred kilometers. Its quietness, high quality and comfort surpassed that of other vehicles in the same class. Six new Golfs produced by FAW-Volkswagen were lined up in an open area on the slopes of the Alps. The team unfurled the Chinese national flag and took a group photo to commemorate the 60th National Day of New China.

Off-road test drive of the Skoda Yeti at Lake Baikal, the largest freshwater lake in the world.

Chapter 9

No Warmth for Mergers and Restructuring

I noticed that there was no auto industry people on the shortlist for CCTV's 2003 People of the Year in the Economy. As a panel member I questioned that, suggesting Miao Wei, General Manager of Dongfeng Motor Group. I argued that the joint venture between Dongfeng and Nissan was the biggest reorganization of a central enterprise that year. Miao Wei, who had led Dongfeng out of trouble, was a master of enterprise re-engineering, and deserved consideration. The panel adopted my suggestion. Miao Wei not only became a candidate, he was also chosen as one of CCTV's People of the Year in the Economy in 2003.

I met Miao Wei in the early 1990s. Later, he followed Lü Fuyuan from the China National Automotive Industry Corporation to become Deputy Director of the Automobile Department in the Ministry of Machinery. We often met during business trips and meetings, and our views on Chinese automobile development made us friends. I was once planning a TV program about car, and often asked him for comments. They were all spontaneous. I would stop him to speak in front of the camera as he was on the way to the cafeteria, and his views were always to the point. We never had to do a re-take.

Dongfeng Motor Co., Ltd., a joint venture between Dongfeng and Nissan, was established in July 2003. The two parties cooperated in trucks, cars, and parts under the co-chairmanship of Miao Wei and Nissan Automobile President Carlos Ghosn. It became China's largest automobile joint venture with the most products, the deepest level of cooperation, and the largest number of employees.

1. Getting out of trouble

Thirty years of challenging topics

Mergers and restructuring were intractable topics in China's auto industry for 30 years.

The chaotic structure of the industry was already being condemned in the 1980s. Ironically, most stand-alone automakers produced less annually than a single production line of a multinational company. These businesses belonged to different central and local departments, and they were from a variety of state-owned, joint venture, and private enterprise backgrounds. Divisions and conflicts of interest, especially pressures of local taxation and costs of production, placed obstacles in the way of mergers and acquisitions across regions.

Mergers and restructuring of the automobile industry didn't happen in isolation. They were part of the wave of globalization, and were driven by technology, markets, and costs.

To observe, learn about, and obtain advanced international automobile manufacturing technology, China's auto industry started joint ventures between state-owned enterprises and multinational corporations in the 1980s. As for markets, the shift eastwards of the center of gravity of the global automotive industry in the last century was itself a process of pursuing markets. When it comes to costs, scaling up for survival is an inborn characteristic of the automobile industry, so costs work like an invisible hand. The costs of developing a car are almost the same all over the world, but the share of design, manufacturing, and component costs can vary greatly between annual production of ten thousand to one million vehicles.

Mergers and restructuring among multinational corporations became a main theme of the global auto industry at the end of the 20th century. The winners lavished tens of billions of dollars to swallow up former star companies, without the slightest civility. Chrysler, the third-largest US automaker, was acquired by Daimler of Germany. The crown of the British auto industry, Rolls-Royce, was carved up by Volkswagen and BMW — one bought the Manchester factory and the other the trademark rights. When France's Renault took over Nissan, the new general manager dismantled the lifetime employment system and laid off employees. When China joined the WTO, that wind blew into the Chinese auto industry.

There were still more than 100 automakers when China joined the WTO in 2001. Most of them were not competitive and were going downhill. The huge losses made some companies into money pits, and they became a burden on local finances. Mergers and restructuring were a difficult first step for getting out of trouble.

Hand-made cars in the Toyota Museum. The museum tells the next generation that manufacturing is a cornerstone of a country's economy.

Restructuring Tianjin-FAW: Prelude to linking up with Toyota

After a period of confusion, the joint restructuring of FAW Tianjin finally took shape with leadership from government organizations. FAW Group and Tianjin Group signed an agreement at the Great Hall of the People on 14 June 2002, six months after China joined the WTO. FAW Group received 50.98 percent of the shares, giving it control of the restructured Tianjin FAW Xiali Automobile Co., Ltd. which was formally incorporated into FAW.

China's auto industry kingpin, FAW Group, was essential to the restructuring. It exerted control over Tianjin Xiali and Huali, prime assets of the under performing Tianjin Group, through acquisition and transfer of state assets and taking on their huge debts.

That gave Tianjin Automobile a chance to recover from its dire situation. It had fallen into an operational quagmire. Monthly sales of the once-popular small Xiali fell to hundreds. FAW, China's first automobile company, had been surpassed by SAIC, losing the position it had occupied in sales and production for ten years. FAW was struggling to create its own brand, and the restructure enabled it to obtain resources for its own economy car. It also helped Tianjin Automobile out of its predicament. However, the merger was of greater significance as the prelude to an international linkup.

How much did it cost to buy a 50.98 percent stake in Xiali and pay off its debt? Both sides kept it secret, but we could be sure it was an astronomical figure. However, the alliance gave FAW Group an underlying choice; to buy a ticket on the mighty Toyota ship, and become bigger and stronger.

At the signing ceremony in the Great Hall of the People, a high-level representative of Tianjin's joint venture partner Toyota sat silently in the back row. After the event, Hattori, Toyota China's chief representative and China expert, revealed that he had been the go-between for the restructuring.

Each of the six major groups which emerged out of the global restructuring had annual production of at least 4 million vehicles. The biggest was over 8.7 million vehicles. By contrast, China's big three, FAW, Dongfeng, and SAIC, which had production capacity of 400,000, were insignificant. After joining the WTO, Chinese auto companies had to go it alone. When the great storms hit them, it was hard to avoid fate.

In this context, FAW's joint venture partner Volkswagen not only appeared weak, but also had less room to maneuver. Toyota, which was mismatched in a joint venture with Tianjin, made quiet overtures to

FAW. This then was the background to the success of the Tianjin-FAW restructure.

FAW General Manager Zhu Yanfeng, Tianjin Chairman Zhang Shitang, and General Manager Lin Yin all had a long-term strategic vision not confined to what they saw before them. The restructure created a profitable result. FAW entered the economy car field, and formed a joint venture with Toyota. In 2004, it became the first domestic automobile group with annual production of more than one million vehicles.

The restructure also gave Tianjin Xiali a chance to regenerate, turning losses to profits in the second year. In 2005, Xiali became the first indigenous brand car company with annual sales exceeding 200,000. However, the good times didn't last long. I felt a bit sad when Tianjin FAW launched the Xiali N5 at the end of 2009, and the FAW badge replaced the Xiali badge. It had been the first to become familiar to ordinary Chinese people. Xiali's innovative R&D progress slowed down, and they faced being marginalized in the competitive small car market. They were at a disadvantage both in product quality and profitability.

Dongfeng Nissan: An attempt at a comprehensive joint venture

The restructuring of Dongfeng's assets began in the 1990s. The company quietly established five car projects: the Fengshen, the Dongfeng Yueda Kia, the Dongfeng Nissan, the Dongfeng Honda, and the Dongfeng Yulong.

In 1992, Dongfeng Group set up a Southern Division, leaving the mountains of northern Hubei. Chen Qingtai had been transferred to Beijing. Ma Yue, who succeeded him as General Manager, gave the Southern Division the mission of being an incubator for strategic development. Young people were recruited, and a new model were adopted — not looking at short-term production but at new developments in ten years' time. Dongfeng sent Deputy General Manager Zhou Wenjie to become General Manager of the Southern Division.

He became known for starting a number of new projects for the Dongfeng Group.

The first project was to cooperate with Ford on its light-duty vehicles. Negotiations began smoothly, but ran aground due to macroeconomic adjustments. The Central Government wasn't approving any new vehicle projects. Zeng Peiyan, Director of the State Planning Commission, suggested that Dongfeng start with parts and components. It looked at sites in Pudong, Zhuhai, Shunde, and Huizhou, before settling on Dayawan Bay in Guangdong for production of engine and chassis components.

Guangzhou Peugeot, which was insolvent, was in the midst of a well publicized restructuring. Although negotiations between Guangzhou Automobile and Opel and Hyundai were well advanced, Dongfeng joined forces with Honda to ask the State Planning Commission if they could participate in the bidding. This made Guangzhou Automobile very uncomfortable, but the State had doubts about its capability then and wanted a big group to participate in the restructuring of Guangzhou Peugeot. So Dongfeng-Honda entered the negotiations and emerged the winner. State Councilor Zou Jiahua personally made the decision to adopt a special mode of cooperation. Guangzhou Automobile and Honda would jointly produce complete vehicles and Dongfeng and Honda would produce engines. The State also adopted this so-called 123 proposal — one joint venture project with Honda, and two companies being established to separately produce complete cars and engines.

Dongfeng Southern Division kept to its original intention of making whole vehicles, and established the Fengshen Company with Taiwan's Yulong, with a stock ratio of 75:25. Fengshen merged with Huadu's Jingan Yunbao, introducing Nissan technology to produce the Nissan Bluebird. This presaged a future joint venture between Dongfeng and Nissan.

In 2000, Dongfeng Yueda Kia restructured. That led to the rapid development of Yueda, a local enterprise based in Yancheng in Jiangsu.

Dongfeng Honda Automobile Company was established in 2005 using Wuhan Wantong All-purpose Light Vehicle as a shell. It started producing the Honda CRV SUV, and captured a large share of a small market, making Dongfeng-Honda the business with the best products and the most profitable under the Dongfeng banner.

In 1997, when it was in trouble during its transition, Miao Wei agreed to serve as Party Committee Secretary and then General Manager of the Dongfeng Group. Dongfeng had made losses year after year. Sales performance was poor for both trucks and cars, and it couldn't even pay wages on time. An independent audit in 1998 showed Dongfeng had a cumulative loss of 540 million yuan, and its survival was in doubt. Miao Wei ordered the Hubei Provincial Party Committee, the Provincial Government, and Dongfeng employees to turn things around in two years. This brought him to worldwide attention.

Under Miao Wei's administration, Dongfeng curbed its slipping efficiency and the cycle of falling car prices in 1999, by taking advantage of increased state orders for military vehicles and debt-to-equity swaps to alleviate market pressure. The Nissan Bluebird and Dongfeng Honda engines became new growth points in the 2000 boom. Dongfeng bounced back. Its car sales, revenue, and profit growth ranked first among the three major groups by the end of 2001, with a profit of 2.5 billion yuan.

I went to Dongfeng in January 2002, and sat in with Miao Wei's leadership team as they gave a report to Yu Zhensheng, the new Provincial Party Secretary. On the basis of that report, I wrote a Xinhua News Agency restricted document — "Dongfeng Motor's Profits Create a 32-year Record." Dongfeng, the old state-owned enterprise, had turned things around and revitalized itself by making use of international asset restructuring. I already knew that a project where Dongfeng and Renault-Nissan Group were allied — the Golden Triangle Project — was pushing ahead.

The joint venture was established in July 2003. Dongfeng and Nissan each held 50 percent of the shares, but the company wasn't called Dongfeng Nissan, rather Dongfeng Limited. Its registered capital was US$2 billion. Dongfeng Motor invested 70 percent of its assets, with phased investment of shares by its subsidiaries and related companies, plus introduction of production and operational entities and its 80,000 employees, into the joint venture. Nissan Motor Co. put in US$1 billion in hard cash.

Nakamura Katsuyuki, President of Dongfeng Limited, cautiously announced his medium-term business plan for 2004–2007 in Beijing on 24 November. Production, sales, and profits would double in four years. By 2007, they would be among the world's top three commercial vehicle companies, and their brand would become one of the most trusted. Miao Wei's goal was to become internationally competitive.

Miao Wei was pragmatic and knew how to rebuild a business. While it was losing so much that it couldn't pay salaries, who would want to spend a billion US dollars to establish a joint venture with Dongfeng? But the old state-owned enterprise made a turnaround under Miao Wei. With annual profit of more than 2 billion yuan, it made a suitable marriage partner for Nissan, Japan's second-largest automaker.

However, benefits had to be mutual for the new Dongfeng to operate smoothly. Nissan was an international company that had also recently turned itself around. To reach production of 1 million vehicles by 2006 — the goal of Ghosn's 180 Plan, Nissan would have to enter China's large market and choose as a partner a strong, successful Chinese company whose products ranged from heavy, medium, and light trucks to sedans.

Partnership negotiations went on for 23 months, involving 2,000 participants at all levels. Many new forms of joint ventures were created in this running-in period, but with everyone working together cooperation progressed smoothly.

Dongfeng Limited's truck products adopted the Dongfeng brand. At the same time, it signed technical cooperation agreements with Nissan and well-known European truck companies to develop new truck cabs and high-powered engines. Dongfeng passenger vehicles adopted the Dongfeng Nissan brand with Huadu in Guangdong and Xiangfan in Hubei as their bases. Dongfeng Nissan stood out for years with its new models.

The restructuring of Dongfeng and Nissan attracted the attention of the State-owned Assets Supervision and Administration Commission

(SASAC). SASAC Director Li Rongrong was as satisfied with it as he was with Kodak's acquisition of Lekai (Lucky Film) — the other international merger he had brokered.

When the joint venture was formed, Dongfeng Group headquarters moved from Shennongjia to Wuhan. But the four Chinese and four foreigners on the management team stayed on at the Shiyan City Dongfeng Truck Production Base in the mountains of northwestern Hubei Province where conditions were still somewhat difficult.

Miao Wei internationalized Dongfeng by separating the main and auxiliary industries and listing the group overseas. Its growth was amazing. Passenger vehicle joint ventures such as Shenlong Automobile, Dongfeng Yueda Kia, Dongfeng Honda Automobile (Wuhan), and the two engine joint ventures of Dongfeng Cummins and Dongfeng Honda all performed well.

Miao Wei left Dongfeng in May 2005 after having worked there for eight years, leaving the Chinese auto industry he had devoted himself to for 27 years. He became Secretary of the Wuhan Municipal Committee, and later became Minister of Industry and Information Technology.

2. Testing the waters

Restructuring SAIC and Nanjing Auto: Having food in your bowl is the most important thing

SAIC and Nanjing Automobile signed a restructure agreement in Beijing on 26 December 2007. That pushed automobile mergers and restructuring in China to a new extreme of size and scope. Chinese people love to talk about the major significance of a thing, but they often overlook its real substance.

I think the key point in this merger was that capital fixed everything. The quality assets of Nanjing Automobile were all absorbed by Shanghai Auto — the stock-market listed company. SAIC Group forked out 2.095 billion yuan of hard cash for it. Yuejin Group took the money and bought 320 million shares from SAIC's investment in Shanghai Auto, making up 25 percent of shares of the newly established Donghua Company.

Nanjing Automobile not only sold for a good price; it even made a profit. At Shanghai Auto's share price of 27 yuan/share on December 27, its 3.2 billion shares gave it a total market value of 8.64 billion yuan, more than double its valuation when SAIC took on Nanjing Automobile's assets. The characteristic Chinese arrangement of maintaining the legal status and taxation channels of the original companies maintained the reputations and interests of both parties.

Free allocation of state-owned assets was abandoned in favor of asset restructuring in accordance with the laws of the market and the norms of listed companies. The dynamics of enterprise development began to play a role.

Nanjing Automobile Factory — the predecessor of Yuejin Group — was older than FAW, having entered Nanjing with the People's Liberation Army when it took the city. It had been a Central Enterprise for nearly 50 years. FAW and Dongfeng still kept the facade of the China National Automotive Industry Corporation when they started anew. There were occasional reports about cooperating with SAIC after they moved to Nanjing in the late 1990s. It seemed the right thing to do at the time. But with conflicting interests and face-saving by corporate decision-makers, talks repeatedly broke down. In particular, their own operational mistakes made the situation worse. Despite the successful acquisition of British MG, they couldn't see any hope of the group making an overall turnaround. But in the first ten years of the 21st century, SAIC, with abundant funds, technology, and talent of Shanghai Volkswagen and Shanghai GM, engaged in indigenous R&D and indigenous brands and rapidly grew to be China's most dynamic car group. Things were no longer as they used to be.

Today, Nanjing Auto, a spinster without a dowry, has married into the vigorous SAIC Group, and has climbed high instead of marrying beneath it. It should be able to live comfortably from now on. Having food in your bowl is the most important thing — for the employees of SAIC and Nanjing Auto, for the shareholders of SAIC, and for the future of China's auto industry.

Hu Maoyuan and Wang Haoliang, representating SAIC and Nanjing Auto, were under pressure to conclude the agreement.

However, they moved deliberately, each considering the future of his own company while allowing for the difficulties of the other side. Their wisdom and generosity were admirable.

Rivalry between the Roewe and MG indigenous brands was a point of cooperation for the Shanghai and Nanjing Automotive Companies. SAIC spent £67 million in 2006 to buy Rover's 75 and 25 models and engine technology. Nanjing Automotive bought Rover's equipment and the MG brand for £53 million. The two successively launched two mid- to high-end models, the Roewe and the MG, and their rivalry began.

At the 2007 Shanghai Auto Show, SAIC Chairman Hu Maoyuan threw an olive branch to Wang Haoliang, Chairman of Nanjing Auto. Hu Maoyuan invited the exhibitors to a banquet that evening and met privately with Wang Haoliang after the meal. They were both far-sighted and the two models having the same technology source was a bond between them. They began investigating full cooperation and integration, and a restructuring took place at the end of the year with the support of central, provincial, and municipal leadership. Their rivalry became friendship.

Nanjing Auto brought the Nanjing Fiat sedan, MG, and the Nanjing Iveco commercial vehicle to SAIC.

The Nanjing Fiat Company, a joint venture between Nanjing Automobile and Fiat, was the source of massive hemorrhaging for Nanjing Automobile Group, with cumulative losses of 2.22 billion yuan. On 21 December 2007, five days before the signing of the SAIC-Nanjing cooperation agreement, Fiat finally agreed to a friendly separation from Nanjing Auto. Volkswagen agreed to transform Nanjing Fiat into a new factory for Shanghai Volkswagen, although they were not convinced it could be accomplished quickly.

SAIC also put up 400 million yuan to repay Nanjing Iveco Commercial Vehicle's debts, augment its cash flow, and give birth to Iveco's new model. Later, SAIC Commercial Vehicle Company, which had heavy trucks, large-scale diesel engines, light trucks, and MPVs, was established within SAIC with Nanjing Iveco at its head. Just after New Year in 2008, SAIC Executive Vice President Chen Zhixin and

Vice President Jiang Zhiwei led their team into Nanjing Auto to begin the 100-day integration.

Shanghai Volkswagen Nanjing Branch, which had been reborn from Nanjing Fiat, started producing the Volkswagen Santana on 1 April. One thousand two hundred of its employees had been trained in Shanghai, in accordance with Shanghai Volkswagen's norms. Shanghai Volkswagen's new equipment and process standards were also implemented. Martin Winterkorn, Chairman of Volkswagen Germany, was surprised when he visited the site. It was rare to see such high-speed transformation of factories, employees, and products. Monthly output was more than 10,000 by October, and local tax revenue increased by nearly 200 million. This had never happened at Nanjing Fiat. SAIC did its work well. No employees were laid off. Employees who used to earn 1,500 yuan a month now earned 3,500 yuan. Nanjing Fiat was revitalized.

Chen Zhixin and Jiang Zhiwei promoted two indigenous brands from Rover — the Roewe and the MG. They unified planning, procurement, research and development, marketing, and manufacturing. This solved MG's problems of over-investment, insufficient sales, and lack of follow-up products.

With unified planning, both brands were produced from the same platform, allowing synergies and economies of scale. Differentiation was accomplished through brand positioning. Roewe, reflecting the elegance of its brand, was positioned in the high-end mainstream car market. MG, with 84 years of history, was propelled by the heritage of its brand. Production at the MG UK factory also resumed after a three year hiatus with the high-end MG-TF sports car.

The SAIC Technology R&D Center and the Nanjing and British R&D branches were mainly based in Shanghai and were linked. A total of 1,800 engineers managed the development of their own brands. Design, products, and craftsmanship took on unified standards using a technical language.

Hu Maoyuan and Wang Haoliang announced the completion of the SAIC-Nanjing integration in December 2010. Nanjing Auto had reduced losses by 40 percent for two consecutive years without a single employee being laid off. The attitude of Nanjing employees had profoundly changed. In the past, they would wait to see what others did. Now they worked as if their lives depended on it. Hu Maoyuan's goal of becoming one family was a reality.

Wang Haoliang, who was infamous for having sold off Nanjing Auto, was smiling at last.

SAIC Group produced and sold more than 3.5 million vehicles in 2010. Nanjing Automobile Group produced and sold nearly 300,000, three times as many as before the restructuring. The Roewe and MG brands achieved production and sales of 160,000 in 2010, laying a solid foundation for indigenous Chinese brands at the middle and high end. During the Twelfth Five-Year Plan period, SAIC invested 10 billion yuan in Nanjing, achieving production capacity of 1 million and annual sales of 100 billion yuan.

The media commented that SAIC had saved Nanjing Auto and Nanjing Auto had helped SAIC. That was a fair assessment.

New Chang'an joins China Aviation to make it into the top four

The Regulations on the Adjustment and Revitalization of the Automobile Industry were promulgated by the General Office of the State Council at the beginning of 2009 as a countermeasure to the global financial crisis. They reiterated the demand for progress in mergers and restructures. As opposed to the previous formulation, the Regulations put Chang'an Automobile on a level with FAW, Dongfeng, and SAIC, and encouraged the large auto companies to carry out mergers and restructures nationwide. They supported companies like BAIC, GAC, Chery, and China National Heavy Duty Truck Group (SINOTRUK) in carrying out regional mergers and restructures.

For the first time in a State Council document, Chang'an Automobile was positioned in the top rank of Chinese automobiles, which undoubtedly enhanced its sense of mission.

FAW, Dongfeng, and SAIC Motor Group sold 1.53 million, 1.32 million, and 1.72 million vehicles, respectively, in 2008. Chang'an

Automobile sold 860,000 vehicles. It had some way to go to catch up with the others.

I had been reporting on military industry departments since the 1980s, and knew how hard it was for the weapons, naval, aviation, aerospace, and nuclear industries to transition from military to civil production. The weapons and aviation industries, because of their strong engine and machining capabilities, had focused development of civilian products on automobiles.

The Southwest Military Industry Bureau invited me to Chongqing in 1987 to report on its heavy-duty truck, the Iron Horse. It was developed by Wangjiang Machinery Factory, which produced anti-aircraft artillery and motorcycles for the Chang'an and Jialing plants. Later, Chang'an brought in Suzuki technology to produce the Chang'an minivan. Director Tan Yanmian took out some pictures and told me they were preparing the Chang'an Alto, which was the first mini-car in China.

Military industry enterprises were classified as automotive industry outsiders, and had difficulty getting approval for sedans, so they started with mini-cars. These survived to become the Songhuajiang and the Changhe. In 2001, before China joined the WTO, annual production of mini-cars was more than 600,000, of which Chang'an, Hafei, and Changhe produced a total of 470,000. Military enterprises held an absolute position in the mini-car industry.

Chang'an rapidly developed, but neither Hafei nor Changhe met expectations. By 2008, Chang'an had sold 860,000 mini-cars, compared to Hafei's 220,000 and less than 110,000 for Changhe.

Just as the State was encouraging the Chang'an Automobile merger, R&D on large aircraft, which had been eagerly awaited for nearly 30 years, fell to the aviation industry. It had no choice but to divest itself of auxiliary industries, shrink its scope of operations, and concentrate on building airplanes.

Xu Bin, General Manager of China Ordnance Equipment Group (also known as China South Industries Group Corporation or CGSC), and Lin Zuoming, General Manager of Aviation Industries of China Group (AVIC), began substantive discussions on restructuring their automotive divisions in February 2009. AVIC Automotive (Aviation Industry of China Automobile Company) was officially listed in Beijing in March. AVIC Group brought Hafei Automobile, Changhe Automobile, and Dongan Power, which produced small automobile engines, into AVIC Automotive. In July, China Southern Industrial Automobile Company, a subsidiary of CGSC, changed its name to Chang'an Automobile Group Company, and was integrated with AVIC Automotive. After National Day, Xu Bin and Lin Zuoming met at a tea house in Beijing where they finalized the details of the restructure.

Government departments gave a green light for the restructure which was in line with national industry policy. The speed of approval was unprecedented.

The two groups reorganized to establish the new China Chang'an Automobile Group Company on 10 November. CGSC held 77 percent and AVIC held 23 percent. Total assets were more than 10 billion yuan.

"It snowed in Beijing today, an omen for a prosperous year. I think it's a good sign." Xu Liuping, newly appointed President of Chang'an, said at the signing ceremony. Changan Automobile Group has since deservedly ranked among the top four in China's auto industry. The new Chang'an owns nine vehicle production bases in Chongqing, Heilongjiang, Jiangxi, Jiangsu, Hebei, Anhui, Shanxi, Guangdong, and Shandong, with annual production capacity of 2.2 million vehicles and engines. Xu Liuping's goal was to forge ahead with a world-class auto company. It achieved sales of 2.6 million vehicles in 2012, and is expected to reach 5 million vehicles in 2020. Chang'an's reach has extended overseas. It has production bases in Malaysia, Vietnam, Iran, Ukraine, and the United States, with projects moving forwards in Mexico and South Africa. Chang'an has established new R&D centers in Beijing and London, following completion of centers in Chongqing, Shanghai, Turin, and Yokohama.

In addition to its existing Ford and Mazda projects, Changan had a new joint venture with an investment of 8.5 billion yuan in Chang'an PSA which would produce the Citroen DS series in Shenzhen.

The Citroen DS was born in the 1950s and was once Citroen's flagship model. It was an avant-garde symbol of innovation and fashion. I went to the launch of Citroen's new DS3 compact car in Paris on 2 February 2009. Citroen said that unlike brands that introduced retro-class classic cars, the DS3 was an example of anti-retro. The design philosophy was not to copy old models, but to create tomorrow's memories with a completely new model. The new DS would be a series. The DS3, DS4, and DS5 would also be launched, comparable to the Citroen C3, C4, and C5. The next day, I drove the DS3 from Paris via Versailles to Orléans for a 150 km test drive. It was stylish and had a floating roof that could be painted in a personalized style. It was powered by a 1.6 liter turbocharged 115 kW engine developed by Citroen and BMW. It excelled in its power and handling.

According to the contract, the joint venture between Chang'an and PSA would be based at a Shenzhen site originally owned by Hafei. It re-activated a production base with a capacity of 100,000 vehicles that had been idle for three years, and promoted Chang'an's sales in southern China. It was the first gift from the reorganized Chang'an.

3. Overseas mergers incur losses

The saga of SAIC's merger with SsangYong: It looked good on paper

When indigenous brands go abroad, they go from exporting cars to building factories overseas, and then participating in international capital operations. SAIC's acquisition of South Korea's SsangYong Motor Company was China's first attempt at an international auto merger. Buffeted by four years of ups and downs, it fell victim to the global financial tsunami.

I think the failure was due mainly to cultural differences, as well as the financial crisis. However, there were problems with SAIC's choice of timing for the merger, the strength of its management team, and its ability to handle crises. Although it did the preparatory work in the early stages, it only looked good on paper. After some flailing around, it failed. But some good came out of it, because it was a stepping stone for the future of the Chinese auto industry.

At the end of 2004, SAIC spent about US$500 million to acquire 48.92 percent of the shares of SsangYong Motors, which was in a precarious state. They increased this to 51.33 percent, becoming the majority shareholder through securities market transactions the following year.

SsangYong was Korea's fifth-largest automaker, producing large SUVs and high-end luxury cars, with a capacity of 200,000. It also had R&D and engineering capabilities.

One of SAIC's purposes with this regional merger was to test the water for building a global management system. SsangYong's SUVs and diesel engines were also strongly complementary with SAIC's product line. With the merger, they could enhance their core competitiveness by bringing into play the product design, development, parts procurement, and marketing networks of both sides.

However, it didn't work. When SAIC took over SsangYong, it learned how wide the gulf was between auto industry cultures in China and South Korea.

Commercial bribery was rampant in South Korea, and the cost of economic crime was extremely low. This resulted in the low capability of the original management team. Suppliers, management, and trade unions had multiple interests. However, when the board of directors dismissed the president, the Chinese didn't have a comprehensive team for international mergers. By contrast, when GM took over South Korea's Daewoo, it immediately selected a 50 person management team from its global organization, with backup support from 500 people. Although SAIC was a leader in the Chinese car industry in its overseas orientation, it lacked international personnel management systems and training.

SAIC made efforts to revive SsangYong.

In 2005, SAIC took over SsangYong, and replaced the former president Soh Jin-Kwan.

In 2006, SAIC management reorganized production systems, and implemented a comprehensive quality control plan.

In 2007, despite the South Korean government's cancellation of diesel vehicle subsidies, the company achieved a turnaround by expanding sales in overseas markets and reducing costs.

In addition, using SAIC's influence, SsangYong successfully carried out financing activities including obtaining large loans and making four bond issues.

In the meantime, Phil Murtaugh, the former CEO of GM China, accepted appointment as Deputy General Manager of SAIC. He spent more than a year at SsangYong. With his overseas management experience, he achieved some success in stemming losses and resolving labor and management relations at SsangYong.

However, the financial crisis in the second half of 2008 broke SsangYong's funding chain and its operations became unsustainable.

Looking back, SAIC people said they hadn't expected the global financial crisis to be so severe, or that SsangYong would be destroyed so quickly. Half of SsangYong's products had been sold in Europe, and now there were almost no sales at all. In Russia, even letters of credit couldn't be opened. Eighty percent of cars were bought with loans in South Korea, but now consumption was sluggish, and banks were reluctant to lend. Unlike Hyundai and Daewoo, SsangYong didn't have its own finance company, so it could only sit and wait for the end to come.

Apart from cooperation in product development, SsangYong was a fully independent Korean company. Therefore, the crisis at SsangYong had little direct impact on SAIC, its major shareholder. As of the end of November 2008, SAIC had a stake of 1.851 billion yuan.

After several rescues, the SsangYong Board of Directors sought bankruptcy protection on 9 January 2009. It took effect on 6 February. The court replaced the board of directors and appointed two Koreans as co-managers. Their task was to develop a rescue plan and notify the shareholders on 22 May. If the court approved the plan, SsangYong would return to normal operations. If the plan was not accepted, it would go into liquidation.

The people of SAIC had thought they were mentally prepared. However, in reality, joining up with a Korean company was harder than they thought, and totally different from their preconceptions about the Korean auto industry.

Korean trade unions were powerful and prepared to go on strike for their share of benefits. As a result, the cost of labor per vehicle at SsangYong Motors comprised 20 percent of total costs, much higher than the average 8 percent in the Korean auto industry. The strength of trade unions was remarkable. More than 100 full-time trade union officials took no part in productive labor but got special vehicles. Management decisions had to be approved by the union. Negotiations and strikes every year brought huge losses to the business.

Strikes occurred three times in the five years after SAIC took over.

The strike in 2006 caused SsangYong a loss of 196 billion won. Coming from an insular culture, Koreans stick together and have strong national self-esteem, but they can also be narrow-minded. The new SsangYong management had planned to amortize the development cost of the S100 small off-road vehicle. In addition to continuing production in South Korea, it planned to assemble some in China, which would increase sales and expand SsangYong CKD exports. That plan was criticized by SsangYong's trade unions for stealing technology and taking jobs away. It was even reported to the Ministry of Justice and a prosecutor issued an injunction against the Chinese management.

After SsangYong went into crisis, the South Koreans were negotiating with SAIC using expressions like "urging," "requesting," and "boycotting." That seemed incongruous with the situation. In the meantime, SsangYong's labor union was playing an excessive role. When SsangYong called for a day of company rebirth, the union designated it as "China Tramples on Korea Day."

The union launched another general strike on 22 May 2009 to oppose a layoff plan. Hundreds of brave union members occupied assembly workshops 3 and 4 and painting workshops 1 and 2 at the Pyeongtaek Plant. They confronted the police for 76 days, finally undergoing a Hollywood-blockbuster style of attack and defense. When more than 200 people had been paid for the costs of their injuries, the union members finally left the workshops voluntarily.

To help SsangYong recover, SAIC promised to use its equity as collateral to raise funds, help find an investor, and maintain SsangYong's sales channels in China. Even so, the union still organized a worker protest at the Chinese Embassy.

SsangYong reportedly asked SAIC, which had been forced to withdraw from management, to keep investing. Cash was king in the financial crisis, and the already cool SAIC had no intention of throwing money into a bottomless pit. SsangYong was taken over in 2010 by Mahendra Group — India's largest tractor manufacturer.

SAAB is in the bag: Beijing Auto keeps a low profile

Xu Heyi, Deputy Director of the Beijing Economic Commission and a metallurgy specialist, took on the job of establishing Beijing Hyundai Motor Company in 2002. In just one year, the Sonata sedan came off the line, fulfilling BAIC's 45 year dream. Xu Heyi became Chairman of Beijing Automotive Holdings in 2006. BAIC sold more than 1.5 million vehicles in just four years, becoming the first state enterprise in Beijing to have a profit of over 10 billion yuan since 1949.

Having made a success of the Beijing Hyundai and Beijing Benz joint ventures, it became BAIC's mission to build an indigenous brand. Xu Heyi, in my view, is like Li Shufu, although their experiences were very different. Both have leaps of thought that ordinary people can hardly keep up with, and a degree of romantic and poetic sensibility. They both turned their eyes towards mergers with famous European and US car brands.

BAIC began its overseas mergers when it competed for Land Rover with Tata of India in 2007. It also tried to acquire Opel and Chrysler. Some of the partners were wary of a merger with a Chinese state-owned enterprise, throwing up one obstacle after another. But that didn't discourage Xu Heyi, it just made him more calculating.

At the beginning of 2009, GM's Swedish SAAB was facing bankruptcy. The high-end car manufacturer which had started as an aircraft manufacturer and was the creator of the turbocharged engine came into Xu Heyi's sights. BAIC used indirect tactics this time, finding a Swedish company — Koenigsegg — to take the lead and act as its buyer.

The plan was established on 24 August 2009. Xu Heyi said it was like "going to sea with borrowed sails." BAIC obtained a 20 percent stake in Koenigsegg through a share swap, and then obtained SAAB technology to build an indigenous brand in Beijing. BAIC General Manager Wang Dazong and BAIC Research Institute Director Gu Lei took part in the negotiations between Koenigsegg and SAAB. Both had worked overseas as high level technology executives at GM, Ford, and other companies for more than ten years before returning to China. The negotiations went smoothly.

As Wang Dazong was boarding a plane from Guangzhou to Beijing on 24 November 2009, just six days before the signing date, he suddenly received a call from the Koenigsegg CEO who told him that the protracted negotiations with SAAB couldn't continue, and Koenigsegg had decided to pull out. They would announce it publicly five hours later.

Wang's head started to spin. The plan that he had promoted for 3 months was coming apart.

Saab Car Museum. Saab started by producing airplanes and aero engines, and then automobiles after World War II.

Participating with Koenigsegg in the comprehensive acquisition of SAAB was a complex transaction. BAIC's aim was clear from the beginning, to go straight for SAAB's current car and engine technology. As for the factory and the brand, Koenigsegg would take them over. Negotiations reached initial consensus, but because of the work involved, the high cost of the merger, and harsh European bank loan conditions, Koenigsegg couldn't handle it and couldn't really afford it, so they had to withdraw.

The plane landed in Beijing three hours later, and the BAIC executives met through the night. Chairman Xu Heyi had the final word. He said, "I think Koenigsegg withdrawing is an opportunity. We'll keep talking with SAAB on the basis of the agreements we've already reached, focusing on acquiring technology."

With SAAB facing bankruptcy or plant closure, GM agreed SAAB and BAIC could negotiate directly. Talks went on day and night. Xu Heyi stayed on in Beijing, Wang Dazong flew to Sweden to manage the negotiations, and Director Gu Lei of the BAIC Research Institute went for days without sleep.

Communications with SAAB in the previous two months weren't wasted. Wang Dazong and Gu Lei, who had studied overseas, found no cultural barriers in their communications with SAAB executives, and they knew what key technologies BAIC needed.

With the bankruptcy deadline getting closer, SAAB and BAIC concluded a draft agreement in three days. GM agreed that BAIC would only buy engines and whole vehicle technology, and they would find another buyer for the brand and plants. Time was even tighter in the remaining three days. On SAAB's side, signatures were needed from the GM President, GM International, GM Europe, and the US Treasury Department. Approvals and bank loans had to be finalized by Xu Heyi in China. As the deadline approached, the negotiating team held their breath until Wang Dazong received a text message saying, "Ha! Ha! The board has passed it!" They all cheered. Things then moved at a pace rarely seen in international mergers. BAIC and SAAB formally signed the purchase agreement for the three SAAB 9-5 and 9-3 variants and complete technology for two turbocharged engines on 6 December. The transaction was completed in Sweden on 11 December. BAIC paid nearly US$200 million. Three and a half tons of documents were loaded onto BAIC trucks, and Gu Lei triumphantly boarded the plane home carrying a hard drive with all the technical files.

The successful deal was praised by SAAB, the Swedish government, and the general public.

SAAB is an acronym for Swedish Aircraft Ab, and because of that aviation origin, the SAAB sedan's campaign slogan was "Born from jets."

To build its own brand and bring back the glory of the Beijing car, BAIC invested US$200 million to buy SAAB's technology but not its factories or trademarks. It avoided the huge risks of acquiring a brand and operating an overseas factory.

Wang Dazong, who had been Vice President of SAIC and had participated in the local development of the Roewe brand, was satisfied that what he had bought was a complete technical system, including digital models, design specifications, processes, and quality standards, as well as a business management system for parts suppliers. It was know-how accumulated from data from millions of vehicles over decades.

"A Beijing opera acrobat clenches his teeth to stop his hat falling off as he turns a somersault. That's what know-how is," Wang Dazong said. By buying a complete system, development time could be shortened by five or six years, and more importantly, with fewer detours. Insiders said just getting the SAAB 84.7 kW twin-turbocharged engine was worth US$200 million.

Xu Heyi told me that three new cars using SAAB technology would be developed and produced at BAIC's new base in Shunyi, coming onto the market in 2012. I asked him if they had bought outdated technology. He answered that it was current production technology and certainly not outdated. He used the twin-turbocharged engine as an example. Getting up to 84.7 kW was quite rare, for an indigenous or a joint venture brand.

4. Low class Geely takes over high class Volvo

A match made in heaven

The 2009 global financial crisis also brought opportunities to the Chinese auto industry. On 28 October, the Ford Company announced that China's Geely Holding Group was the preferred bidder for the Volvo Car Corporation. Geely and Ford signed a framework agreement for the acquisition of Volvo on 23 December. Ford didn't plan to keep any share of Volvo. Geely acquired it in full, and Li Shufu was impressive again.

The industry and public opinion had been unfavorable towards the acquisition, and joked that it was just a show with no real substance. And no wonder. Could it be a match made in heaven when a young, grassroots Chinese company wanted to acquire a century-old European luxury brand? Ford had invested US$6.45 billion to buy Volvo in 1999, and was ready to sell it off for US$1.5 billion. Could Li Shufu revive it? There were also problems with Volvo's trade unions, for which there was the example of SAIC's acquisition of South Korea's Ssangyong.

Acquiring Volvo was a prudent step for Geely. At the beginning of the project, Li Shufu said to me privately in Hangzhou, "The plan to acquire Volvo was put forward in the company in 2002. I very much wanted to acquire a good reputation by purchasing a famous international brand. I wanted to change the bad image of indigenous brands for relying on imitations and replicas. We didn't want to just go to the Third World countries, we also want to take our place among European and US markets with solid technology."

In the previous two years, Geely had taken control of British Manganese Bronze Holdings (MBH) which produced the classic London TX4 taxi, and had successfully acquired the Australian DSI automatic transmission plant through international capital markets. Geely's international capital operations were maturing.

In 2010, it entered virgin territory by acquiring Volvo. Geely and Ford signed an agreement in Stockholm on 28 March 2010, and Geely finalized it in London on 2 August. It sealed the deal at 1 pm local time, with US$1.3 billion in cash plus US$200 million in a note. Geely became China's first multinational auto company, ranking in the top 500 global businesses.

The reason behind Volvo's past losses was that as a subsidiary of Ford, its marketing and product strategies were completely tied to Ford's global strategy. When Ford had problems, it had no time to deal with Volvo. The key to revitalizing Volvo was to get it back to itself.

Li Shufu said, "Geely was Geely, and Volvo was Volvo. Geely made a commitment that Volvo would adhere to its core values of safety, quality, environmental protection, and modern Nordic design. Volvo's current Swedish and Belgian factories, R&D centers, dealer networks, procurement channels as well as trade union agreements would be retained, and emerging markets including China would be actively explored. Volvo would be headed by an independent management team headquartered in Gothenburg, Sweden."

Li Shufu didn't see that Geely had to integrate with Volvo. "It was a tree, well planted where it was. You could water it and give it fertilizer, but there was no need to strengthen it," he said.

Geely's recovery plan was to build a new Volvo base in China with production capacity of 300,000 vehicles, nearly doubling its output. It had turned things around by 2011. Within five years, the luxury Volvo brand had global sales of 1 million vehicles.

Geely Group Chairman Li Shufu and Ford Motor Company CFO Lewis Booth shook hands in Gothenburg, Sweden on 28 March 2010 when Geely acquired Volvo.

Shanghai Geely Zhaoyuan International Investment Company, registered in Jiading District, Shanghai, was the principal entity for Geely's acquisition of Volvo. Geely paid Ford the $1.3 billion through it. Geely Zhaoyuan is a joint venture established by Geely Group, the state-owned Daqing Assets Supervision and Administration Commission, and Shanghai Jia'erwo Investment Company, in a ratio of 51:37:12.

Joining Volvo

There was a slate blue sky and rich greenery outside the window as I flew to the Volvo headquarters in Gothenburg on 16 September 2010. The first directors meeting since Geely acquired Volvo had just ended that morning. We had lunch with Chairman Li Shufu and President and CEO Stefan Jacoby.

Volvo came out of loss and started to make profit globally in the first half of 2010. Volvo began this journey to excellence amid the hopes and fears of the global auto industry and markets, but on firm ground with magnificent prospects and a realistic vision.

I had known Li Shufu for more than ten years. He was no longer the image of persistence, restlessness, and outspokenness. In the great game of globalization, his wisdom and courage, strategic thinking, and cautious advance, had met with greater respect. But his basic truthfulness, simplicity, and fairness hadn't changed.

He told me privately, "I'm not out to make profit, but to bring Volvo back to the global status of BMW and Mercedes-Benz as a luxury brand."

"We will have huge growth. The people at Volvo have been too conservative and don't act until they are 200 percent sure. I act when I'm 80 percent sure. When those conservatives work together with a progressive like me for two or three years, you'll see a change," he said. That reflected his philosophy.

"Can Volvo technology be transferred to Geely's local production systems?" I asked. Li Shufu replied simply, "No, we can't put them together. One is a high-end product, the other is popular, and the technical systems are different. Costs will go up if you try it."

Li Shufu was only permitted to enter Volvo for the first time after the acquisition was done. Safety is a gene Volvo was born with. At the Volvo Museum, the words of founders Gabrielsson and Larson at the 1927 launch of the first generation of Volvo cars still ring true: "Cars are driven by people, so the guiding principle for everything we do must be safety first." Today, Volvo has become synonymous with safety.

Since 1959, Volvo has invented a number of safety devices such as three-point seat belts, anti-lock braking systems (ABS), and airbags, and is the leader in the field. It was the first to claim that new Volvo cars could achieve zero collisions and zero casualties by 2020.

We observed a head-on collision test between an S60 and a V70 at the Volvo Safety Center. Thanks to a series of protective measures including the front bumpers, the engines and cabs of both cars were effectively protected and no fuel leaked out, verifying their high standard of safety.

Volvo is also a pioneer in the field of environmental protection. Its first environmental regulations were developed as early as 1972. Volvo's three-way catalytic converter with oxygen sensors was launched in the US in 1976. Now all emissions controls are based on it. Volvo also achieved a recycling rate of 85 percent in 2002.

We visited Volvo's Torslanda plant in Gothenburg, which was built in 1964. It has 2,300 employees and female workers make up 27 percent — the highest proportion in the world. The Volvo V70, S80, XC70, and XC90 are produced there. The assembly line builds to order, producing a vehicle in 73 seconds.

Volvo also opened a design center and wind tunnel laboratory for the Chinese, conducting test drives for all models including hybrid and electric vehicles.

Everyone I spoke to at Volvo expressed genuine acceptance of the acquisition by Geely. From top levels to ordinary workers, they all had admiration for Chairman Li. There was a theme in what they said, "Geely and Ford are different. Geely does not see Volvo as a junior. Li

Shufu respects Volvo and its culture very much and strives to restore its former glory. Volvo has a good investor."

Lynk & Co: Creating a new high-end brand

When Li Shufu took over Volvo, he restored the profitability of the luxury brand. He created global sales of 570,000 vehicles in 2017, setting records four years running. That was a surprise for the people who had laughed at him. Moreover, with the success of Volvo, Geely Group jumped from being China's grassroots to becoming a world-famous car company and one of China's most successful indigenous brands.

Geely and Volvo jointly established the completely new Lynk & Co brand in 2017. That shot, aimed directly at Volkswagen's and Toyota's successful product competition in China, realized Li Shufu's original intention in acquiring Volvo.

China Euro Vehicle Technology (CEVT), Geely's European R&D center, was established in Gothenburg in 2012. Geely decided to use Volvo's advanced technology and understanding of new directions in global markets to launch a conceptually innovative global car to compete with mainstream European and US brands. It would be manufactured simultaneously in China and Europe, with joint procurement, and separate sales.

I went to the launches of the Lynk & Co 01 in Berlin in September 2017, the Lynk & Co 02 in Amsterdam in April 2018, and the Lynk & Co 03 in Tokyo in September the same year. The press conferences were creative and international, with the global media present. I also visited the CEVT Center in Gothenburg, and the Zhangjiakou plant and Ghent factory in Belgium where the Lynk & Co would be produced alongside Volvos.

The cars I saw embodied Volvo's safety and environmental concepts, durable quality, minimalist Scandinavian design and people-oriented technology.

An Conghui, CEO of Geely Automobile and Lynk & Co, is an outstanding manager who joined the company when it started. The mid- to high-end Borui sedan and Boyue SUV he developed completely changed Geely's image. He declared that taking the new Lynk & Co up a level could only be accomplished by getting ahead of the competition and targeting the technical standards and user experience of luxury models like Audi and Lexus.

Three markets would be addressed. Volvo, under the Geely Group banner, would mainly target the luxury market. Lynk & Co would target the joint venture market. Geely would focus on the indigenous market.

The strength of Lynk & Co is the Compact Modular Architecture (CMA) platform jointly developed by Volvo and Geely. The world's newest small and medium-sized car platform shared by Volvo and Geely products is the most advanced in the world in terms of physical and electronic performance. Lynk & Co has become global, adapting simultaneously to European, US, and Chinese markets. Development testing is done according to the most exacting standards.

The IT industry's involvement in cars has created transformational pressures for traditional cars. However, An Conghui said, "Lynk & Co is born in the internet era on a blank piece of paper and there has been no transformation. Our starting point is to realize the full integration of the traditional car and internet thinking. We didn't go to extremes. It's first and foremost a car, safe, reliable, clean, and smart, and in practice is no less than that. It will also be a smarter, more convenient, and more powerful mobile terminal than a mobile phone. It can be continuously upgraded at no cost."

The company also has an innovative sales model. Orders are made through a single network ordering platform, whether they are placed by mobile phone or in a 4S shop. There are no price increases, and no one can jump the queue. Orders are locked in by a data delivery program. The 4S stores are responsible for integrating social media resources, increasing consumer awareness, deliveries, and after-sales service. That is very different from the traditional way of selling cars.

Perhaps it was the success of the product promotion, but on the first day of online sales, 6,000 orders for the Lynk & Co were cleared in just 137 seconds, taking 57 minutes to complete payments. It was a sales miracle for an indigenous brand.

Geely is producing both capital and technology as it reaches a high point. It has a voice in the global market with its mastery of the most advanced core technologies.

Lynk & Co, a high-end brand, uses the CMA basic modular architecture jointly developed by Geely and Volvo. The Lynk & Co 02 SUV features high-performance control, intelligent interconnection, and passenger and pedestrian safety.

05 VW's comeback: An impression of Winfried Vahland

Winfried Vahland was President and CEO of Volkswagen China Investment Corporation. He had light-colored hair, a sturdy build, and was charismatic. Two mornings a week, before going to work, he would join a group of young people to play football at the British school behind the company. They used to play hard; once, he injured a tendon in his leg when he was tackled by one of his staff. He had to use a cane when he traveled and went to work.

I never admit defeat

Born in 1957, Vahland had an iron will. His Chinese joint venture partners and his subordinates all feared him, hated him, cursed him behind his back, and were swept along by him. But at the end of his five-year term, people saw a Volkswagen China that had climbed up from the bottom of a steep ravine.

When he stepped down, Beijing, Shanghai, and Changchun all held grand farewell parties with champagne and fireworks for him. Everyone was happy with the success.

When Vahland came to Beijing in 2005, Volkswagen was like a mighty ship having a hard time navigating China's waterways. He spoke with me and a few media colleagues not long after he took charge. I told him bluntly that when Volkswagen monopolized half of the market in the 1990s, it had bragged it was unbeatable. Most of Volkswagen's models in China were old, but it ignored price competition and believed that only rich people bought cars. Volkswagen's decision-making was too slow as it faced the market surge when China joined the WTO. It also couldn't breathe easily as it was being chased by two joint venture competitors — US and Japanese/Korean, each going their own way in terms of cost control and market sales.

Vahland responded to my criticism by saying, "Mr. Li, you are too polite. It's been hard for me to accept Volkswagen's market share going from 50 percent to 17 percent. Carlos Ghosn's success at Nissan was due to a 20 percent cost reduction over three years. It took them that long. However, FAW Volkswagen and Shanghai Volkswagen need even greater cost reduction. I have been in contact with FAW and SAIC leaders to come up with ways to stop the decline and grow our market share. We have to stop thinking and start acting. Every department and every process has to be involved. I brought together 50 German and 150 Chinese engineers. We discussed every component from morning till night for a week and came up with a plan to reduce costs."

Vahland soon launched Volkswagen's famous Olympic Plan. Its signature goal was to reduce overall costs by 40 percent by the end of 2008. Ten to twelve new products would also be launched every year, and Volkswagen would become the most competitive brand after 2008. Its products would be more than 80 percent localized. Vahland said to me, "I never admit defeat. To maintain Volkswagen's leading position I'll spend what I have to, but we won't lower quality by even a millimeter."

By the next year, Volkswagen's sales had increased by 24.3 percent, setting a record for its 22 years in China.

Successes of the Turbo era

Three new strategies were introduced — the Power-train Strategy, the 2018 Strategy, and the Southern Strategy — with tangible results. Sales more than doubled during Vahland's five-year tenure, reaching 1.4 million in 2009. Market share remained stable at around 18 percent. It maintained its leading position and profits increased substantially.

The Power-train Strategy is especially worth mentioning. Safety is emphasized in German cars. They are heavy, and their fuel consumption has long been higher than that of Japanese cars. Vahland was determined to improve performance in terms of energy saving and emission reduction, and that was achieved in all models in 2010. Fuel consumption and emissions were brought down by 20 percent. Through Vahland's persistent efforts at Wolfsburg, Volkswagen brought the world's most advanced Turbo Fuel Stratified Injection (TFSI) and the Direct Shift Gearbox (DSG) technology to China. In order to make the new technology available locally, Volkswagen soon built TFSI and DSG factories in Dalian and Shanghai. The Dalian plant began producing 150,000 DSG units annually in 2010. Vahland planned a two phase expansion of the plant to produce 600,000 units just two months later.

Vahland brought a Turbo era to the Chinese car industry, setting technical benchmarks which competitors had to follow. This included engine downsizing and Volkswagen's BlueMotion clean energy, guaranteeing a 20 percent reduction in fuel consumption and emissions by 2010.

Vahland was relentless in achieving his goals. He went back to Wolfsburg and clamored for new technology and resources. He also used personal relationships to convince the Chinese decision-makers in the joint venture to introduce new models. He had a unique vision, and all the models he backed became popular.

One thing demonstrated Vahland's personal role. Even models produced at the new Volkswagen plant in the US didn't use advanced technologies such as in-cylinder direct injection and dual clutch.

Five years prior, most Chinese consumers had preferred Japanese cars. Toyota and Honda, with their fine interiors, consistent quality, and low fuel consumption, were their first choice. By the end of Vahland's term, Volkswagen represented a German car which was ten years ahead in technology. It was recognized by car enthusiasts for its meticulous production skill and its fuel-saving and safety technologies.

Volkswagen's Chinese comeback couldn't have happened without Winfried Vahland.

Calling him the Pharaoh behind his back

The German pronunciation of Vahland is closer to "Fa-lan-duh", and initially Chinese employees in the company secretly called him the "Pharaoh" (Fa Lao). Eventually, everyone called him the Pharaoh behind his back.

Vahland was unconventional and far-sighted. He didn't just think ahead a year or two, but ten or twenty years. However, he paid attention to details and reviewed sales reports every day.

Yang Meihong, Vice President of Volkswagen China Investment Corporation, took part in the formulation of the Volkswagen China 2018 Strategy and felt strongly about it. She told me that long-term planning was not something Vahland took lightly. He held meetings with each department and analyzed their targets. They had to update equipment, change production processes, develop product planning, build networks, and train personnel. He would have her coordinate with each department in advance and he personally helped solve problems. Detailed targets were finalized, and participants had to sign off before leaving. Long-term strategy was part of the daily work. Signing off meant something now.

Vahland was a quick thinker, and it was hard for his subordinates to keep up. Working with him was exhausting. You needed to know what was happening and how to deal with problems. He was angry if you didn't. Both Wolfsburg headquarters and Chinese partners had to be present at closed-door meetings, and no one could leave until the problem was solved. But everybody soon adapted to that high-intensity activity. When Yang Meihong was Volkswagen China Public Relations Director, Vahland often sent her to meetings outside her business scope. That puzzled her at first, but she realized Vahland wanted everyone to have an integrated understanding so that they could help him make decisions.

My children

However, Vahland was open to people's opinions. I was the first to propose differentiation of the two joint ventures. I recall speaking about it with Vahland soon after he became CEO of Volkswagen China in 2005. I said that FAW-Volkswagen could build traditional Volkswagens, while Shanghai Volkswagen understood local development. I specifically mentioned my Shanghai Style concept. I said, "Any foreign car that comes to Shanghai, whether it's from VW or GM, regardless of its production or marketing, will be imprinted with the Shanghai Style."

It was difficult for Vahland, a newly arrived foreigner, to understand the cultural heritage and reputation of Shanghai industrial products in Chinese markets over the past hundred years. He said Volkswagen should treat everyone equally. However, he began to differentiate between the two when he spoke with the media.

Vahland was effective in localization during his tenure. In the Olympic Plan, it was decided to develop half of all new products in China.

On an evening with a light spring breeze in 2008, Vahland invited the Chinese management of the two joint ventures and a few media friends to see the Lavida and the new Bora developed by the joint venture. The two models were distinctly different. The new Bora focused on the original bold style of Volkswagens in Germany, while the LaVida had a graceful Shanghai style. Vahland affectionately called the two new cars "my children."

With a glass of champagne in his hand, Vahland pulled me aside and asked me which car I liked. I told him I liked the LaVida because it was designed like a Volkswagen but incorporated Chinese concepts of beauty. When I asked him which one he liked, he said the new Bora. I smiled and said that Germans still seemed to prefer real Volkswagens.

New pride in Skoda

Volkswagen is a multi-brand car group with luxury brands such as Bugatti, Lamborghini, Bentley, and Audi, and production cars such as Volkswagen, Skoda, and SEAT. Vahland brought Skoda to China and it became another rising star for Shanghai Volkswagen.

In a hotel surrounded by coconut palms in Sanya, Hainan, Vahland told me that was where he negotiated with Shanghai Volkswagen's Chen Zhixin. Vahland was still Vice President of Skoda then. They talked for two days without a breakthrough. Chen Zhixin suggested going to a large, airy room upstairs facing the sea. That's where they reached the agreement to jointly produce the Skoda sedan.

Interestingly, Vahland's next position when he left China was as Chairman of Skoda. He made a lot of changes there. He brought back product planning, and even changed the design of Skoda's badge. That was Vahland.

Skoda let him keep in contact with China. He came to Shanghai to meet Zhang Hailiang, the new General Manager of Shanghai Volkswagen, in December 2010, and they met with the media to announce Skoda's new China strategy and new production arrangements. He was proud that Skoda had been in China for three years, and had grown at more than 60 percent a year. China had become Skoda's largest market. I went to Lake Baikal in Russia that summer to test-drive the Skoda Yeti SUV. It was a lot of fun off-road. I saw Vahland at the Paris Motor Show and suggested the Yeti be introduced to China as soon as possible, and it definitely shouldn't be lengthened. Vahland announced that the Yeti would be put into production at Shanghai Volkswagen in 2013. However, I heard Zhang Hailiang say that lengthening it would make it more suited to China.

At the end of the meeting, Vahland came over and said to me, "Li, please come to Skoda next year. I want you to see the new models Skoda hasn't shown outsiders yet."

I looked forward to that day.

Chapter 10

Opportunities for China in the Crisis

On 16 September 2008, GM celebrated its 100th birthday amid the haze enveloping the viewing gallery at the top of the 492-meter-high Shanghai World Financial Center, just after the tower's completion. The height made guests dizzy. GM Global CEO Rick Wagoner's speech was shown on the big screen. The centenary was celebrated simultaneously at venues in the US, Germany and Mexico.

The celebration was an expression of hope in adversity. The financial storm triggered by the US subprime mortgage crisis in 2008 had spread rapidly through the US, Europe, and the rest of the world. A few years before, the global economy was still engaged in a bubble-generating carnival. Financial derivatives, the Internet, and real estate seemed to bring endless wealth. However, the collapse of several financial institutions on Wall Street caused uproar. Wherever the shock wave went, banks went bankrupt, factories laid off staff, imports and exports shrank, and business withered. It became an economic crisis sweeping the world, reaching a climax in 2009.

1. An unexpected global crisis

GM topples after its centenary

Just after its 100th birthday, GM bore the brunt of the global financial turmoil. Negotiations with unions broke down, and its restructuring plan was rejected by President Obama. CEO Wagoner was forced to resign on 30 March 2009, and a debt-for-equity swap was rejected in

May. By June all rescue plans had failed, and GM declared bankruptcy and collapsed in its 101st year.

GM's headquarters at the Renaissance Center in Detroit. Just after celebrating its 100th birthday, it suddenly faced bankruptcy in the financial crisis.

It was called the darkest day for US industry.

When you are big, your problems are big. The old company was carrying a lot of weight. GM employees earned an average of US$70 an hour, 1.75 times as much as their Japanese counterparts. The company supported almost a million retirees. Employee medical expenses were as high as US$2,400 per car. The workers had a strong union, the United Automobile Workers (UAW), which never gave an inch on employee welfare. GM couldn't keep carrying those costs. There was a big fall in sales when the crisis came. It lost US$15.3 billion in 2008. By 2009, with government transfusions of US$25 billion, there was still no sign of improvement, and it went on to lose US$6 billion in the first quarter.

The crunch had come, and there was no way to turn things around. That was the last straw for the Obama administration. It forced Wagoner out, then announced GM was in bankruptcy and had been taken over by the US and Canadian governments. The US Federal

Government bought 61 percent of the shares, and Canada 10 percent. Debts owed to unions and creditors were swapped for equity, and GM became the world's largest government owned enterprise.

Industry insiders said the company was behind the times, and didn't developed new products. It had been satisfied making large, fuel-inefficient vehicles like SUVs, which made it noncompetitive. I think they were unfair.

Smith and Wagoner had been drastically reforming GM to fix its chronic problems since the 1990s. The first was getting out of the dilemma of multiple independent brands. Wagoner announced a new brand strategy in 2004: to develop a global product platform meeting the needs of different brands and markets through a modular architecture that reduced development costs. By the end of 2008, Shanghai GM had the latest Chevrolet Cruz, Buick Regal, and LaCrosse models on GM's global platform. Unluckily, the financial crisis was too much for it, and GM collapsed before the new cars came out.

GM was committed to developing new energy vehicles. Their development of pure electric vehicles, plug-in hybrids, and hydrogen fuel cell vehicles had been impressive since the 1990s. However, in 2003, GM proposed a grand plan to redefine the car in five to seven years. Perhaps, on reflection, using new energy to redefine the car and change its traditional form was too far ahead of its time.

After the global tide for electric cars ebbed in 2003, GM developed the plug-in hybrid Volt, which was called an extended range electric vehicle. It didn't push for increased cruising range like other electric vehicles, rather it had a choice of gasoline and electric options. It was a breakthrough that showed electric vehicles were feasible. Although the Volt was launched in 2008, it still needed three years of technical refinement, and got off to a slow start. Unfortunately, it couldn't save GM.

There wasn't a transformation of the global energy structure to support these efforts. GMs difficulties increased as its billions of dollars of investments failed to succeed. Time waits for no man, and it made Wagoner a tragic figure.

GM's bankruptcy was a milestone. By the end of 2010, despite constantly replacing its CEO, GM finally unburdened itself and divested its non-performing assets. It got $49.5 billion from the US government, and recovered much of its vitality. But the Obama administration wasn't interested in a government-owned enterprise. GM returned to the New York Stock Exchange on 18 November, raising a total of $23.1 billion, making it the largest public offering in history. Signs of new life were just around the corner.

Toyota stumbles

"When you get to the foot of the mountain, the road appears. Where there is a road, there is a Toyota." Toyota's reputation for quality and reliability took root in the Chinese market 30 years ago. Even today, most people who buy Toyota cars go for the brand.

I've had a lot to do with Toyota since the 1980s, and I've always thought it was reliable. It was the first company we went to for help to reform our auto industry. Chinese auto companies had regarded Toyota as a standard of quality management for 30 years. It was only interested in selling cars in China in the early years. FAW and SAIC expected the introduction of technology in their joint ventures, but that was ignored.

Akio Toyoda, a former race car driver, sitting on the roof of a newly launched Toyota at the 2011 Shanghai International Auto Show.

Older generation automakers are still indignant when they remember how arrogant Toyota was.

Toyota's overseas expansion focused first on the US, then on Europe. In China, a virgin territory for the car industry, Volkswagen of Germany, PSA of France, and Chrysler and GM of the US had been pioneering for many years. They had turned countless somersaults, but still couldn't say they had gained a foothold. Toyota arrived late, when the market had finally matured, and had a reputation for selling imported cars. Its boast that it had no rivals seems empty today.

In 2005 Katsuaki Watanabe, who replaced Fujio Cho as President of Toyota, was determined to be bigger than GM and become the global leader. Watanabe, known as the "cost killer," saved Toyota billions of dollars by asking employees to "squeeze the last drop of water out of the towel." But simply reducing costs couldn't change the fate of an over-expanded Toyota. Impacted by the global financial crisis, it made its first annual loss in 70 years in 2008. Watanabe took responsibility and resigned.

In June 2009, Akio Toyoda took over the reins of the endangered company, becoming its eleventh president. In a speech he said, "I'm sounding the horn to tell everyone the car is driving near the edge of a cliff." Sure enough, Toyota's recall broke out in the US three months later.

It was a defining event in the 70 years since Toyota was founded. The damage to the Toyota brand couldn't be overestimated.

A friend from Toyota China came to my home on Lunar New Year's Eve 2010 to let me know about Toyota's global recall. I said that anything that Toyota was going to do would be closing the door after the horse had bolted. Even when Toyota recalled 8.5 million vehicles worldwide because of accelerator pedal and braking problems, new problems kept appearing. It had fallen into a deep well, and still hadn't reached the bottom.

Sure enough, during Spring Festival, accidental acceleration of Toyota cars caused the deaths of 34 people in the US, and two more problematic Toyotas were recalled. The US Congress called Akio Toyoda to a Toyota Quality Crisis hearing on 24 February, where he apologized and offered compensation. After that, he came to apologize to China.

I don't think the US government was holding Toyota back to benefit its domestic auto industry. Considerable numbers of problematic Ford and GM cars had been recalled in the past. In the interests of Toyota's suppliers and its US employees, the US government had chosen to say nothing for more than a year. In the end, Toyota's secrecy about thousands of quality complaints and fatal accidents worsened the situation.

It takes more than one cold day for three feet of ice to form, and the unprecedented recall was the culmination of a series of problems with Toyota's manufacturing and operating concepts. Toyota had spent too much effort becoming the world's top producer. Its pursuit of cost reduction and profit maximization was at the root of the crisis.

Tests on new Toyota cars were mostly carried out on indoor stands and on racetracks. European cars carried out over 3 million kilometers, sometimes up to 8 million kilometers, of on road testing. This could include the extremely cold Arctic, the dry Sahara, and high humidity tropical rain forest. For example, the accelerator pedal problem and frost forming in the brakes would have been discovered much earlier with long-distance testing.

Success with energy-saving and emissions reduction, such as the new hybrid technology, must be acknowledged. But because costs were kept down, the gap widened with European rivals in traditional technology and products. While both the Volkswagen Golf and the Toyota Corolla were international best-sellers with sales of more than 26 million each, the 10th generation Corolla was at least five years behind the sixth generation Golf in terms of engine, gearbox, chassis, and suspension.

The recall involving many markets around the world was a bitter pill to swallow, and Toyota had to find a radical strategy to prevent such problems from recurring. Chinese auto companies should draw lessons from Toyota's stumble.

Toyota's sales didn't decrease, but reached a record high of 846,000 in 2010 on the back of a boom in China's auto market. The contributions of its two Chinese joint venture partners were important. Feng

Xingya, General Manager of GAC Toyota, and Tian Cong, General Manager of FAW Toyota Sales, were old hands who had been in the Chinese auto business for many years. They dealt with the crisis calmly, adjusting their product structure to consumer demand and strengthening second and third-line sales networks. Their phenomenal sales went against the market. Toyota President Akio Toyoda was moved by the support shown by Chinese partners and consumers when the company was in danger.

Toyota has a tradition of doggedness. In the spring of 1936, shortly after the launch of the Toyota G1 truck, one of the trucks loaded with goods broke down on the main road at Hamana Lake. Toyota quickly sent a vehicle to carry the goods for them. Toyota Kiichiro, the 42-year-old founder of the company, regained customers' trust by rushing to the scene and getting under the truck to find the problem.

The crisis was an opportunity for Toyota to look at itself. After 70 years, it had some of the illnesses of big business. Former Toyota China General Manager Kato once told me that when he joined Toyota 33 years ago, if a worker at the beginning of the production line was too busy, his companions would come and help. The team leaders also treated young people like younger brothers. Now that the company has become more globalized, the rank levels are more distinct. Losing the principles of the previous generation might have caused the crisis. Akio Toyoda often said Toyota should go back to its old principles. Every department had to find where it had deviated from the old principles and make changes. Do sales people still know every customer's face like they used to?

I interviewed Toyoda twice, at the 2011 Shanghai Auto Show and at the Anchi Club in Toyota City in Japan. I asked what the old principles were. He said, "It means a return to the original intentions of Grandfather Toyota Saki when he invented the automatic loom: to free the weaver from hard manual labor, to produce high-quality products in a modern way, and to make society wealthy. Many of the older generation have told me it will bring hope back to Toyota. In the future, Toyota must create good cars that give consumers a feeling of value."

Toyota had just released its China strategy. People were concerned about whether Toyota was serious about localizing in China. Akio Toyoda said, "We will establish a research and development center at Changshu in Jiangsu. The biggest change will be to take some developmental functions previously performed in Japan to Changshu and establish one of the world's largest test sites there. Toyota sells cars all over the world. We need to provide the most suitable car for local consumption by bringing together the needs of consumers in different regions. We have to develop cars that are resource-friendly and friendly to the global environment, while striving to occupying 15 percent of the Chinese market."

What kept Volkswagen steady?

The Volkswagen Group's 2008 annual report was issued in March 2009. Chairman Martin Winterkorn announced in Wolfsburg that Volkswagen achieved retail volume of 6.23 million. This was higher than the previous year even in the difficult market environment of 2008. In the global financial turmoil, Volkswagen became the only large group with overall growth in sales and profits.

Volkswagen's ability to withstand the financial storm wasn't accidental. Volkswagen, representing engineering culture, went opposite to GM's and Toyota's well-publicized development of new energy vehicles. It had been dedicated to the core business of the traditional automobile industry for more than a decade. With the development of turbocharged small and efficient engines, clean diesel technology, lightweight bodies, and dual-clutch gearboxes, it followed mainstream energy saving and emission reduction trends with downsizing and low energy consumption.

In the 1980s, Volkswagen Chairman Carl Hahn's strategy was to expand the market and the size of the group. During his term he brought SEAT and Skoda into Volkswagen, and expanded cooperation with China.

Hahn's successor, Ferdinand Piech, who came from a family with a long history in automobiles, was typical of someone who had worked at Volkswagen for nearly 20 years. Since 1988, he had served as

Chairman of Audi, Chairman of the Volkswagen Group, and Chairman of the Volkswagen Board of Supervisors.

Piech was mad about technology. A friend told me he had seen Piech at a banquet. While the others were chatting, his head was bent over a mechanism he was sketching on a paper napkin. During his time at Audi, when he was faced with the challenge from Japanese cars, he worried that the European auto industry was losing its way and worked hard to promote innovation in traditional technologies.

He introduced technologies like turbocharging, the Quattro full-time all-wheel drive, and the lightweight tubular aluminum body. Audi still holds a leading position in the global automotive industry.

In the ten years Piech was chairman, Volkswagen acquired top luxury brands like Bugatti, Lamborghini, and Bentley, and developed the Volkswagen flagship luxury car, the Phaeton and a high-powered 16-cylinder engine. At the other extreme, he presided over the development of the Lupo, the world's most fuel-efficient car which could go 100 kilometers using three liters of fuel. On the day he left the chairmanship, he launched a concept car with fuel consumption of less than one liter per 100 kilometers. On the other hand, he wasn't keen on capital operations in financial markets. He was always criticized for the increased costs from pursuing technology. However, Volkswagen had the most brands and the most solid technology of any multinational group because of those measures.

Martin Winterkorn, Volkswagen Chairman after 2006, continued to implement Piech's ideas. He kept up the return to core business, improved the platform strategy and modularization, sped up product development cycles, and maintained a leading position in innovation in traditional technology. Volkswagen pioneered Turbo Fuel Stratified Injection (TFSI) and the Direct Shift Gearbox (DSG), leading competitors to follow suit. Efficient green power, which includes electric vehicles as well as its modular strategy, enabled Volkswagen to maintain competitive momentum.

Volkswagen, with its solid technical foundation, uniquely maintained steady growth and vitality during the global financial crisis.

In July 2009, after a rumor that it would be taken over by Porsche, Volkswagen announced a surprise move. It would turn the tables by buying Porsche in full at a price of 8 billion euros. Porsche, the world's top luxury brand, would become Volkswagen's tenth brand. When the restructure was done, Volkswagen would fulfill its long-cherished wish to outdo Toyota in sales and profits, and become a global flag-bearer.

As the first multinational company to enter China, Volkswagen played an important role setting high standards for the Chinese car industry. The Chinese market also brought a generous return. In 2010, Audi, Volkswagen, and Skoda — the three brands of Volkswagen's two joint ventures in China — couldn't keep up with demand.

Intelligent, glass-walled parking structure at Volkswagen's delivery center in Wolfsburg, Germany.

Fiat-Chrysler's reorganization: A classic case of technology for capital

Sergio Marchionne, CEO of the newly formed Fiat-Chrysler Group, arrived at his new office in the Chrysler Building in Detroit on 10 June

2009. That was just 42 days before President Obama announced Chrysler was going into bankruptcy. That set off a new round of technology-for-capital restructuring with some smaller players taking over the majors in the global auto industry.

Marchionne, whose background was in finance, took charge of the nearly bankrupt Fiat Auto Company in 2006. I was the first Chinese reporter to interview him. He carried out drastic reforms right away. The first was withdrawing from the marriage with GM the same year, taking a $2 billion breakup payment. Next, the Fiat GrandPunto, Bravo, Linea, and other small cars and energy-efficient engines came out. The launch of the 1.4T engine and the rebirth of the Fiat 500 accorded with world trends in energy savings and emissions reduction reshaped the "King of Small Cars" and brought Fiat Auto into profit.

The financial turmoil still hadn't become fully apparent in autumn 2008, and Marchionne predicted that the auto industry would face the same market as the Wal-Mart supermarket chain, and people who made luxury consumer goods would inevitably get into trouble. The good times were over for cars, and people in high positions who were not up to the task would have to go. He also predicted that after mergers, there might be five or six large groups left. Annual sales would have to reach 5.5 to 6 million vehicles to make a profit. Marchionne modestly positioned Fiat as a new force in Europe that would be expanded by absorbing medium-sized groups.

The unexpected financial crisis brought Fiat's dreams closer to reality.

Chrysler went into bankruptcy at the end of April 2009 and the US Supreme Court approved Fiat's restructuring plan for it. Fiat would hold a 20 percent stake in the new Chrysler company, valued at US$8–10 billion. In exchange, Fiat provided Chrysler with a small car platform that could be produced in the US, without asking a penny in cash. Fiat would also provide energy-efficient engines and access to Fiat distribution networks as bargaining chips for a further 15 percent increase in its stake in Chrysler, to be carried out in three steps.

Marchionne was wise in offering Fiat's advanced technology, car models, management, power train, and distribution networks, which

Chrysler needed to save itself, in exchange for equity rather than for cash. That showed the value of Fiat's technological innovation. Warren Buffett had money, but couldn't save Chrysler.

Fiat's wisdom and knowledge couldn't be grasped with traditional thinking. Marchionne said, "The world we live is changing. We won't be able to copy the past any more, so we have to keep innovating."

A year after the restructure, Marchionne silenced those who ridiculed him for buying a losing lottery ticket. Fiat's share in Chrysler increased from 20 percent to 25 percent in January 2011. Marchionne said it would reach 50 percent. We were watching a snake swallowing an elephant.

With China's huge market prospects, it was important for Fiat, already divorced from Shanghai-Nanjing Cooperation, to find another partner. Guangzhou Auto Company (GAC) came to an agreement for Fiat to provide technical support for the independent brand small cars and engines it was developing. It would help GAC build its own brand with Fiat's mid-to-high-end Alfa-Romeo sedan platform technology and advanced engines. A new Guangzhou-Fiat joint venture car company appeared in 2010.

Chrysler, creator of the Jeep, SUVs, and MPVs, celebrated the 70th anniversary of its Jeep brand in April 2011, and decided to try the Chinese market again. After several changes, Zheng Jie, a true Shanghai woman, was appointed General Manager of Chrysler China Sales Company at the end of 2010.

Zheng Jie was a journalist from *China Daily*. In 2002, she was appointed a director of GM, becoming the first Chinese to be a director of such a large company. She was known for her professionalism, commitment, and perfectionism. It was hoped the Shanghai-savvy Zheng Jie would rewrite the story for Fiat-Chrysler.

When this spectacle appeared before us, it was accompanied by an unprecedented global financial storm. GM and Toyota were veterans which had made outstanding contributions to the global industry. However, wrong strategic choices had been devastating to them.

Volkswagen and Fiat focused on optimizing the power train and smartly taking restructuring opportunities as they came. There were no dazzling tricks; they just took up market initiatives in the turbulent waves of the financial tsunami.

This turbulence weighed heavily on Marchionne, who died suddenly of cancer in July 2018.

2. A unique landscape

A fire in the cold winter

In the second half of 2008, China was still recovering from the excitement of the Beijing Olympic Games. As the clouds of the global financial crisis loomed over the Chinese economy, the car market was kept warm by the enthusiasm for the Torch Relay, after which it suddenly fell into a trough. The celebrations for automobile production reaching 10 million, already put on hold, were canceled.

The industry was filled with pessimism at the beginning of 2009. Most economists predicted zero or negative annual growth in the auto market.

Before Spring Festival, I took part in a CCTV discussion of "How Chinese cars will make it through winter." When the host asked about the domestic market, Li Shufu, the guest of honor, surprised everyone by saying: "I don't feel cold. It's been a warm winter for us at Geely. We are working overtime, and there are a lot of opportunities for us."

Li Shufu's unlikely predictions often came true.

After four months, the Chinese auto market stopped falling and rebounded. By the end of September, global auto production was still bleak. In the first three quarters, production fell by 22.9 percent year-on-year globally, 30.5 percent in Europe, and 41.5 percent in North America. The Chinese market was the only bright spot, growing by nearly 46 percent. That offset the impact of the global financial crisis and the domestic economic downturn, and created the second sales boom of the 21st century.

In October 2009, the 10 millionth car produced in China that year came off the Changchun FAW assembly line, fulfilling the dream of several generations of Chinese automakers. China surpassed the US with 13.64 million cars produced and 13.79 million sold at the end of the year, becoming the world's number 1.

Shanghai GM's new Sail was launched in Chengdu in December 2009. Shanghai GM wasn't dragged down by its parent company. Sales increased by nearly 80 percent that year. In the past, the GM Group's Asia Pacific center had been Australia, and later South Korea. Now it had moved to China.

With disaster close by, Shanghai GM was unharmed. It was highly independent. GM's bankruptcy didn't stop Shanghai GM going it alone, even proposing to bring the Buick brand into SAIC or as a joint venture. That year, Shanghai GM got through the test with its best performance ever.

Shanghai GM General Manager Ding Lei told me that the crisis in GM was making it possible to export joint venture products. Shanghai GM's current models came from GM's global platform. In other words, if it was too costly for GM products to be produced in a certain country, the factory could be closed down and they could be exported from

China. The cost of the models developed by GM's Shanghai Pan Asia R&D Center was low. They were more competitive and a large number of exports were expected. It was likely the Sail would go onto the global market through GM Chevrolet's sales system.

The Chevrolet Sail was launched at Yantai in October 2010.

SAIC Vice President Ye Yongming took over as General Manager of Shanghai General Motors in February 2011. The Chinese car market slid sharply that year, but Shanghai GM launched six new models, including the Chevrolet Camaro and Aveo, which were in strong demand. Ye Yongming said, "Ten years ago, Shanghai GM sold GM models in Shanghai. But now, GM is enhancing its global competitiveness with models developed in Shanghai."

It was a watershed day when the 10 millionth vehicle that year — a Red Liberation J6 truck — came off the production line in Changchun in October 2009. The Chinese auto industry, which had existed for more than 50 years, had finally joined the ten million annual production club. By the end of the year, 13.79 million cars had been sold in China, 48 percent more than in the previous year. China overtook the US in production and sales. That contributed to China's economic recovery and stable economic growth. And Chinese cars have since occupied the top global position for 20 years.

Growth in the automobile market was a result of the Automotive Industry Adjustment and Revitalization Plan issued by the State Council at the beginning of the year. A series of incentives encouraged automobile consumption. Tax on small-displacement vehicles under 1.6 liters was halved, cars came to the countryside, and ways were found to replace cars which didn't meet pollution or fuel efficiency standards.

It was the first time in 50 years that the government had encouraged consumption of vehicles, especially sedans.

The auto market boom continued throughout the 2009 global crisis, due to those policies, and allowing more freedom was a win for the government. Chinese people and Chinese companies are tough, and they know what hard work is. Nurture small seedlings and they grow into trees.

The arrival of the age of car consumption in the first decade of the new century was a direct cause of market growth. Per capita GDP increased from US$1,000 in 2000 to US$3,700 in 2009. Living standards and purchasing power rose significantly. Cars went into large numbers of urban and rural households, providing strong support for the market. According to a rule of the world automobile market, when the number of passenger cars per thousand people reaches more than 20, there will be a period of rapid growth. China had reached an average of 25 vehicles and grew rapidly. Demand remained strong, with some fluctuations, for the next 15 years.

The number of cars reached 86 per thousand in several large cities, while in smaller third tier cities it was only eight. The total population of second- and third-tier cities was around 800 million. As long as the economy maintained 8–9 percent growth, the vast inland cities would support China's auto industry for many years. This was proved by Shanghai GM's Wuling, Changan, and other mini-car companies which had sales of more than 100 million annually.

The localization war among luxury brands

Germany's Der Spiegel reported in June 2010 that German auto manufacturers were still stuck in the economic crisis in the second half of 2009, and the German government's scrappage policy was having little effect. However, strong demand for Mercedes-Benz, BMW, and Audi cars in overseas markets became a lifeline for these luxury brands. To meet demand in the US, and especially in China, Mercedes-Benz, BMW, and Audi stepped up production, some factories even giving up summer holidays.

Luxury cars drove the Chinese auto market in the post-crisis period. Sales of luxury cars exceeded 700,000 in 2010, making up more than 6 percent of the market. They had 10 percent of the global market, and 25 percent of the German market, so there was still room for growth.

Germany's three luxury brands tried hard to compete in the Chinese market. Other luxury brands such as Volvo, Cadillac, Lexus, and Jaguar Land Rover also did well.

In 2010, Audi produced 225,600 vehicles, BMW 168,900, and Mercedes-Benz 147,600. However, the ranking was reversed in terms of growth: Mercedes-Benz 146 percent, BMW 117 percent, and Audi 109 percent. It's hard to know which one wouble be the winner.

The China strategies of the three big German companies in the post-crisis era were all about localization, unlike Mercedes-Benz and BMW which had been keen on imports in the past.

Audi's CEO went to Changchun on 22 October, to celebrate FAW-Volkswagen selling its 1 millionth Audi in China. The Audi 100 had pioneered the production of luxury cars in China starting with a technology transfer 22 years earlier. China was already Audi's largest overseas market due to localization of the full value chain. Audi's next goal was to sell 1 million more Audis in the next three years.

Mercedes-Benz president Cai Che came to Beijing two days later to host a ceremony for the lighting of the three-pointed star emblem on the Mercedes-Benz building. He announced that Mercedes-Benz would integrate more fully into the local culture in China, investing 3 billion euros in its production base in the next few years, including an engine factory. At the same time, Mercedes-Benz would increase the proportion of domestically produced vehicles from the current 30 percent to 70 percent over the next three years.

BMW had already joined the battle for luxury brand localization. The BMW 5 Series in 2008 was an extension of early attempts at localization by BMW China President Christof Stark, a fluent Chinese speaker. By the time Brilliance Auto's BMW production came to full capacity, a second BMW plant was being built. BMW announced that its production capacity in China would reach 300,000, and localization of parts would reach 60 percent.

The author and his son Li Man test drove the new Range Rover Velar in Norway in spring 2017. It was expected to be produced in China.

3. Crisis brings opportunities for China

The confidence brought about by ten years of WTO membership

Ten years after China joined the WTO, Chinese cars were becoming more globally integrated, and the technology, capital, restructuring, innovation, market, talents, and opportunities which were obtained were exceeding expectations. Those ten years gave Chinese automobiles confidence to control their own destiny and to go against the market in the crisis.

Zhang Xiaoyu, Vice President of China Machinery Industry Federation, who had taken part in the reform of China's auto industry, said the confidence brought to the industry by ten years of WTO membership was why Chinese autos could beat the crisis.

"I had been involved in GATT negotiations, which later became WTO negotiations, since 1992," Zhang Xiaoyu recalled. "The theme was protecting the Chinese auto industry. In the negotiation process, Chinese industry was divided into three categories. For category A, including agriculture, the automobile industry, and the finance industry, the challenges were greater than the opportunities. For category B, including the machine tool and machinery industries, opportunities and challenges coexisted. For category C, including the textile industry, opportunities were greater than challenges. In the , the actual situation was the other way round, not at all what I had thought at the time."

Shanghai GM assembly line workers.

Zhang Xiaoyu said, "From 1994 to 2000, during the Zhu Rongji administration, the automobile industry slowed, and there was almost no growth. We had reached 1.2 million vehicles in 1992, but it was not until 2000 that we reached 2 million vehicles. That took eight years. It was impossible to get there in 1998 and 1999. The whole industry suffered losses, and FAW and SAW couldn't pay wages. It was the same situation we had before China joined the WTO."

"It was difficult in 2001, when the national economy kept being adjusted. China's car industry could only produce a few hundred thousand vehicles. It was worrying because the lambs could see the wolves getting closer. The plan we gave to the Central Committee stressed that the car industry had to be protected for at least eight years after joining the WTO. Import tariffs shouldn't be less than 30 percent. Finally, there would be a transition period of six years, with tariffs reduced year by year to 25 percent."

"China made a commitment to the WTO to open up its markets. US$6 billion worth of automobiles, including parts and components, was imported the year we joined, increasing 15 percent annually until the quota ended in 2006. As a result, imports increased by five times to reach US$30 billion in 2008. Imports of complete vehicles grew from 40,000 to 400,000 a year, greatly exceeding our commitment. However, exports of Chinese cars grew even faster, from US$20 billion to US$46 billion in 2008."

Zhang Xiaoyu also said, "Openness and global competition after we joined the WTO brought about large investments that were unthinkable in the past. From 2001 to 2005, fixed asset investment in the automotive industry was 235 billion yuan, equal to total investment of the last four five-year plans. Annual investment has been more than 100 billion yuan since 2006, and was over 600 billion yuan in 2010, more than the full amount of the previous 25 years. Ninety percent came from foreign capital, corporate profits, and stock markets, and state investment was less than 1 percent."

In the ten years after China joined the WTO, the auto industry stopped being problematic and became inspiring. Beginning from 2000, China's annual automobile production was 2 million, including 600,000 sedans. That reached 18 million by 2010, including 10 million sedans. In the past decade, vehicle production increased nine-fold (sedans 16 times), and the average annual growth was 24 percent, and it kept growing.

It kept growing by more than 20 percent for ten years. That had never happened before. Even countries like Japan and South Korea only had five years of fast growth.

Before we joined the WTO, state-owned automobile enterprises stood apart and access to capital outside the industry was restricted.

That's why Li Shufu, trying to get permission to build automobiles, said at least give him a chance to fail. Markets opened up after 2001, and private enterprise and international brands came in. The so-called "big dogs" (central enterprises), "puppies" (local state-owned enterprises), and "wild dogs" (private enterprises) were brought in.

According to data provided by the 2010 Automotive Blue Book jointly compiled by the Development Research Center of the State Council and the China Automotive Engineering Society, the industry's operational income was nearly 3 trillion yuan in 2009, total profit was 200 billion yuan, and more than 300 billion yuan was paid in taxes. The Chinese auto industry, which had worried that the WTO would wipe it out, was making a lot of money.

Occupying the heights of technological innovation

Will the crisis give China more influence?

Diplomat Wu Jianmin said, "When the crisis settles down, the new international order of the 21st century will take shape. This crisis is important for China. Confronting it depends on technological innovation. Throughout human history crises have brought forward new technologies, inventions, and breakthroughs. A crisis is an opportunity. Many countries addressed the crisis by considering the immediate future as well as the long-term. The US Government gave money to save the three major US auto companies, but also demanded that they reshape themselves to occupy the commanding heights of automobile technology in the 21st century. My view is that the financial crisis will advance society through three major revolutions: an energy revolution, an industrial revolution, and a lifestyle revolution. This is also a good time for China. If we transform successfully, we might occupy a favorable position in the new global structure. But if we don't do well, then it will go badly for us, and it's hard to say where we shall stand."

Attractive opportunities to learn and to innovate came with international mergers and acquisitions. With parts, vehicle, or design companies, multinational mergers extended upstream into the automotive industry chain, and the pace of China's integration of global automotive resources became more urgent.

General Motors in the US came out from 16 months of bankruptcy protection in November 2010. The "China factor" played a role in its return to Wall Street. Its joint venture in China kept growing and was a highlight of GM's re-listing. In the new GM CEO and CFO's "road show," they emphasized their success in growth markets like China.

In GM's global offering, SAIC spent US$500 million to acquire a 0.97 percent stake. SAIC's participation in GM was more symbolic than real. It marked the two sides becoming partners after more than a decade of hard work.

After 30 years of joint ventures and cooperation, and after ten years of global competition, China became the undisputed world's largest automotive nation, and the fledgling Chinese industry accelerated the overall pace of globalization.

Zhang Xiaoyu and Wu Jianmin had passed away, but their insights still shone a light on the path for China's auto industry.

Great Wall Motor's Haval Vision 2025 concept car is equipped with 5G, automatic driving mode, remote control, and facial recognition functions. This smart car, which can think and communicate with the driver, will come out in 2023.

06 A missed lunch: An Impression of Phil Murtaugh

Phil Murtaugh, an American, was short, stocky, and clever. At just over fifty, his hair was already going gray. He worked for GM in the US and overseas for 32 years, becoming the President of GM China. He later served as Vice President of SAIC and President of Chrysler Asia Pacific. Despite all this, he had none of a big boss's arrogance or haughtiness.

He was kind and open-minded. He said he first met me at the Detroit Auto Show in 1997. GM Asia-Pacific President Rudy Schlais pointed me out to him and said, "That guy, Li knows a lot about Chinese cars." Since then we became good friends, slapping each other on the back when we met up, although there were times when we disagreed. Murtaugh told other people that I spoke sharply and didn't listen, but he valued my opinions.

Products can be developed indigenously in big markets

Chinese officials had given Murtaugh numerous honors, like the Magnolia Award and a Shanghai Green Card, which he was very proud of. He was straightforward and easy to get along with. When he was a member of the negotiating team in the Shanghai project in the 1990s, and later when he served as Shanghai GM's Vice President, he got along well with his Chinese counterparts Hu Maoyuan and Chen Hong.

"Seeing someone else's point of view" was a phrase he often used. When there were differences of opinion, Murtaugh wanted to understand what the other party was thinking. "Resolving contradictions is not about winning and losing, it can serve both sides. When you divide a cake, one gets more and other gets less. But if we both make the cake bigger, both get more," he said.

We would meet up every once in a while and had a Sichuan meal which he liked. They weren't interviews, we would just chat. I remember going to his office in 2002, the second year after China joined the WTO. We ended up talking for two hours and went without lunch. However, the notes I made are still relevant.

I asked him what differences there were in pricing structures in China and the US.

He told me that for Buicks produced in Shanghai, apart from costs and profit, there was also 17 percent value-added tax and 3–8 percent consumption tax; consumers had to pay another 10 percent purchase tax; plus 30 percent import tariffs on spare parts and so on. When consumers buy a car, 40–45 percent of the cost was in taxes. Take the Buick GS as an example. In China, taxes and fees alone were US$20,000. In the US, the federal government tax on auto products was 4 percent; state taxes range from 0–10 percent, which meant that you needed to pay taxes of 4–14 percent to buy a car. To expand car consumption in China this would have to change.

I asked him what GM's earnings were used for. The media often thought that auto manufacturers' profits gave them room to lower their prices.

Murtaugh said companies didn't usually disclose their profitability. As a reference, average profitability in the global automotive industry was 3 percent. Of course, Shanghai GM's profitability was more than that. Its profits were mainly used for capital reinvestment, new technology, R&D for new models, retirement funds and employee welfare funds, as well as shareholder dividends. If profits decreased, one of the direct results was a weakening of product development, and without new products to satisfy the market, companies would lose customers.

I said many people were worried China would become the world's factory for multinational car companies, with core technology developed abroad, and China simply a manufacturing center.

Murtaugh said, "That possibility exists in theory, but I don't think it will really happen in China. Because the market is so big, customer demand is different from the rest of the world."

For example, the environment and driving conditions were very different, he continued. It's almost impossible for a product designed and developed outside China to be completely suitable for China. That's why GM wanted to establish a Pan-Asian R&D center in Shanghai. The same was true of other markets around the world. GM's Detroit headquarters has a strong R&D capability, but there are also R&D centers at Opel in Germany, Holden in Australia, and in Brazil. At the outset, technical improvements and localization would take place, and eventually, indigenous development capabilities would be formed.

He said, "There's little probability China will become a pure production center for foreign auto companies because everyone agrees China is one of the largest markets. My experience is that large markets can support large R&D centers to develop products indigenously."

One day in spring 2004, I received a call from GM's PR Director Ms. Zheng Jie, to tell me Murtaugh had suddenly resigned as CEO of GM China. He had worked in China for almost ten years.

Everyone wondered why he left, but he and GM seemed to have made a deal beforehand which they stayed quiet about. I heard he had disagreed with his bosses in Detroit and had refused to compromise, leading to an impasse. In the ten years he was in China, Murtaugh opened up China's huge market for GM; he also did a lot of good for the global integration of China's auto industry.

Don't resort to ideology

He was asked to join SAIC soon after he left GM. SAIC had just acquired Ssangyong, its first involvement in overseas operations, and was having difficulties dealing with its Korean labor unions. Murtaugh was appointed Executive Vice President of SAIC's overseas business, also serving as Vice Chairman of Ssangyong.

We met when he came to Beijing in the first half of 2004. When I asked him if Ssangyong would make a turnaround the following year, he shrugged. "I don't know, it depends how lucky we are," he said. "They say in the US you can only see something clearly when you're within 20 yards of it."

He said, "SAIC took on a burden when it bought Ssangyong. It isn't that they weren't smart. The benefits were apparent when they bought the business off-the-shelf. There were products, networks, and talent. The problem is they didn't do proper due diligence. There were history and side issues they didn't find out about until they went in."

Murtaugh continued, "When SAIC took over the product line, the Korean people said it was technology outflow. Everyone knew that was not the case, it was just a political excuse. There had been no problem when Hyundai and Daewoo models were produced in China. As soon as Ssangyong signed a technology transfer license with SAIC for the Xiangyu KYRON model, the Korean press and tabloids ganged up on it, and the government passed an Anti-Technology Outflow Act. Unlike Americans, South Koreans are strongly nationalistic and that causes problems for Americans or anyone else who invests there. SAIC learned from its Ssangyong experience and avoided all sorts of problems by doing thorough due diligence when it bought Rover."

I said it's possible that other Chinese companies had encountered problems with overseas acquisitions because of ideological reasons.

Murtaugh disagreed and said, "Don't put setbacks down to ideology. When US companies first came into China 20 years ago, they used politics as an excuse when they began losing money. By the mid-1990s, they had learned about the Chinese market and had become successful, which they attributed to their own good strategy. That was saving face, not politics. China's economic growth is so fast that foreigners are naturally skeptical and fearful. The key thing is to have a car that is competitive in the local market, and price is by no means the only factor. China's auto industry can't force international markets to fall into line with it. It will only succeed by understanding new markets and catering to them."

Murtaugh said, "In fact, Chinese cars encountered more problems overseas than the Japanese did. Japan is the only country to have invaded the US, traumatizing generations of Americans. Japanese cars went in and out of the US market in the 1970s, but they persisted with market research to understand what their customers needed, and relentlessly opened up the market with fuel-efficient products, low prices, and good service. Americans were open to this and interested in the benefits, so they finally came around."

Ssangyong became profitable while Murtaugh was in charge.

Plain speaking doesn't always get the result you want

After Daimler-Chrysler split up, Murtaugh went to Chrysler as President of Chrysler Asia Pacific. The first problems he faced were Chrysler's joint ventures with BAIC and Chery.

He investigated thoroughly and determined that the culture, technical standards, and quality management of the two companies weren't up to Chrysler's standards. He decided to withdraw from negotiations with the two Chinese companies.

BAIC and Chery objected, thinking Murtaugh was making things difficult for them, but their cultures were in fact different. He told me privately that Chery would have to replace all its management if it wanted to improve. While he sounded merciless, I understood Murtaugh. When it comes down to it, you can't only compete with technology and capital. It's also a contest of ideas.

Murtaugh was cursed for this in China. Chrysler also thought he hadn't shown enough initiative and he had to move on again. Looking back, his judgment was sound. Ending the joint ventures was good for Chrysler, as well as for BAIC and Chery. Chrysler went into bankruptcy during the financial crisis. It wouldn't have been able to survive if it still had the joint ventures.

I often think Murtaugh's career would have gone more smoothly if he had been more diplomatic, but plain speaking is needed, and I respect him for that.

Chapter 11

New Energy: Don't Hang on a Single Tree

The first time I visited the Mercedes factory in Stuttgart was in 1995. It was impossible for me as a Chinese to imagine that before constructing the new factory on undeveloped land, the company would have entomologists coordinate with lighting designers so that there would be no damage to the environment of local insects at night. The outdoor illumination was configured so it wouldn't disturb the insects' breeding environment.

To reduce energy consumption, shutters and blinds were computer controlled for the best ventilation and light. Rainwater was collected for watering the shrubbery.

In a workshop, I was curious about a square stainless steel tray placed under a water pump. The technician told me it was to prevent condensation from dripping directly onto the floor. I wondered why it mattered. My escort replied that if any harmful substances were dissolved in the water droplets, they could collect in the cement over time. When it's time to demolish the factory, the discarded cement blocks could pollute the environment. Therefore, the water needed to be collected and processed centrally. This gave me an insight about protecting the environment for our children and grandchildren.

This was my first lesson in environmental protection and energy savings.

1. A road map for new energy

Responding to issues

There would be no cars without petroleum. They are a product of the 20th century oil age. One of the worst things about cars, however, is that they consume a lot of gasoline.

Automobiles grew along with a surge in global oil production at the beginning of the 20th century. Two global oil crises, in the 1970s and the 1990s, didn't slow down the global automotive industry. People looked for technology to fuel cars with less gasoline. The average fuel consumption of today's cars is only about one-third that of 30 years ago. Gasoline which could supply 10 million cars then, can meet the demand from 30 million cars today. The EU and US plan to halve automobile fuel consumption in the next 30 years.

China was oil-poor in the first half of the 20th century. Kerosene, gasoline, and diesel fuel for everything from lighting to driving almost all came from abroad, from the US before Liberation (1949), and from the Soviet Union after that. Imports were cut off when China and the Soviet Union fell out in 1961. The older generation still vividly remembers that buses on Chang'an Avenue were fueled by huge bags of coal gas on the roofs. China stopped being oil-poor when large oil fields such as Daqing were discovered in the 1960s and 1970s. China was closed off and the oil crises had little impact.

China became more dependent on oil as its economy took off in the 1980s. It changed from an oil exporter to an oil importer. The surge in global oil prices and the nightmare of imminent oil depletion began to disturb the Chinese economy.

In the early 1990s, I kept hearing the prophecy that the world's oil resources would be nearly exhausted within 30 years. I took it seriously at the time. Today, new oil fields are being discovered, oil production technology keeps improving, and the world's proven reserves can be exploited for at least 50 years.

By 2020, an estimated 1.1 billion vehicles will be traveling on roads around the world and it will be difficult to rely solely on oil. One third of China's oil is consumed by automobiles now. 250 million tons of oil will be required by 2020. Finding new energy sources is an urgent matter. Tomorrow's cars will use a variety of energy sources including fuel oil, natural gas, and shale oil, as well as renewable energy such as bio-energy, nuclear, hydrogen, wind, and solar.

Emissions pollution is another bad thing about cars.

Like steam trains which used to belch smoke from their smoke-stacks, exhaust pipes of early cars also emitted black smoke. This was carbon particles which were not fully combusted, as well as carbon monoxide, sulfur dioxide, lead, nitrogen, and hydrogen compounds that were invisible to the eye and harmful to the environment and the human body.

However, the global automotive industry has achieved remarkable results since control of vehicle exhaust emissions began in the 1960s. California was the first to issue a series of regulations. In the 1970s, Europe and Japan also formulated regulations to control automobile pollution. These regulations became more restrictive, promoting con-tinual technological innovation. By the year 2000, emissions of carbon monoxide, hydrocarbons, and nitrogen-hydrogen compounds had fallen by 96 percent, 96 percent, and 76 percent, respectively, in the US; by 95 percent, 96 percent, and 92 percent in Japan; and by 85 percent and 78 percent (for each of the latter two) in Europe.

There has been an unprecedented rise in the world's environmental awareness since the early 2000s, especially regarding atmospheric warming caused by carbon emissions, the greenhouse effect, and the crisis facing human survival. Even though carbon dioxide emissions from cars wasn't on the list of things restricted for directly harming the human body, the EU and the US proposed that by 2020 average carbon emissions per kilometer per car shouldn't exceed a strict 95 grams.

China has also stepped up regulations. It implemented an Environmental Protection Law, an Air Pollution Prevention and Control Law, and air pollution quality standards equivalent to the European Union's Euro 1 to Euro 5 vehicle exhaust emission standards.

It's more than 130 years since Mercedes Benz invented the first car in 1886. Automotive technology is now undergoing revolutionary changes including new energy, advanced power systems, smart inter-connections, and new materials.

A New Energy Road Map for the 21st century

After countless twists and turns, the global auto industry reached a consensus on a road map for new energy technology around 2010.

I was shown the 21st Century New Automotive Energy Technology Road Map. It describes both near term and future new energy technologies, which can be developed in parallel:

1. High-efficiency internal combustion engines and bio-fuels such as bio-diesel and ethanol;
2. Hybrid electric power;
3. Plug-in hybrid vehicles (also called extended-range electric vehicles);
4. Pure electric cars (its icon is a short line, meaning it's only suitable for short and medium distances and cannot completely replace traditional cars);
5. Hydrogen fuel cell vehicles. Hydrogen energy is plentiful, has zero emissions, and is the ultimate solution for new energy sources.

In 2014, the author drove from Milan to Pisa with Dr. Dieter Zetsche, Chairman of the Mercedes-Benz Board of Directors, and talked about new energy vehicles.

Electric propulsion runs through every mode involved in the new energy road map. All attempts at energy savings and emissions reduction using new energy will be accomplished by electrical propulsion. Broadly speaking, all new energy vehicles are electric vehicles. That is often kept deliberately vague in China.

2. Optimization is our best choice

Combustion vehicle technology is still the main battlefield for energy savings and emissions reduction

Even by the most optimistic estimates, pure electric vehicles will not exceed 50 percent of total global sales by 2030. Most of the 70 million vehicles produced globally every year have internal combustion engines. In 2017, China was still the top producer of electric vehicles. But the 580,000 electric vehicles sold only made up 2.5 percent of domestic passenger car sales.

Technological innovation of traditional internal combustion engines is mainstream to energy conservation and emissions reduction. Turbo Fuel Stratified Injection (TFSI) greatly enhances full combustion of gasoline. Energy loss from shifting gears is almost zero with the invention of the Direct Shift Gearbox (DSG).

Energy savings and emissions reduction are complex systems engineering projects involving aerodynamics, weight reduction, tire friction, smart electronics, and energy management.

Bench tests show that if the vehicle drag coefficient is reduced from 0.36 to 0.32, roughly 10 percent, the fuel consumption can be reduced by 0.15 liters per 100 kilometers; in unimpeded driving, the fuel consumption per 100 kilometers can be reduced by 1 liter or more. The Mercedes-Benz E-Class Coupe has become the world's most aerodynamic mass-produced car with a drag coefficient of 0.24. Volvo invested EUR 20 million to upgrade its wind tunnel laboratory in Gothenburg, Sweden, to continue to reduce the drag coefficient of its new cars.

Reducing the weight of the car body is more than just skimping on materials. High-strength aluminum alloys are increasingly replacing heavier steel. Audi established a lightweight body laboratory nearly 30 years ago. With its all-aluminum body technology, the company is far ahead of the rest of the industry. For every 100 kilograms of vehicle weight reduction, fuel consumption per 100 kilometers can be lowered by 0.3 liters. With electric vehicles, endurance increases by 7.5 percent. An Audi car with this technology can be more than 150 kg lighter than others in its class. It is the leader among luxury brands.

The leg we have to stand on

The China Automotive Engineering Society invited me to the 2009 Huadu Auto Forum. The forum was of great interest because of an argument between two authoritative figures, He Guangyuan and Long Yongtu. I was one of three guests who were invited to make a keynote speech. I proposed for the first time that we have to keep up with the world's leaders in traditional automotive technology despite the national enthusiasm for electric vehicles.

I stressed that development of new energy vehicles and optimization of traditional vehicles are two legs of the Chinese auto industry, and we cannot do without either. Research and development on new energy vehicles commands the heights of tomorrow's technology. It's the leg we have to step forward with. Optimizing traditional vehicles will get instant results for energy conservation and emissions reduction. It's the leg we have to stand on.

Government led enthusiasm for electric vehicles have undoubtedly left innovation on traditional automobile technology out in the cold. In the first 413 million yuan tranche of funding for the National 863 Energy Conservation and New Energy Vehicles project, R&D funding for energy-saving and environmentally-friendly internal combustion engines was only 16 million yuan.

Officials have pointed out that it would be foolish to set a timetable for completely banning internal combustion automobiles. It would be better to optimize petroleum engines first, so as to be on the same timetable as the world's best.

Energy savings and emission reductions gained from optimizing internal combustion engines are greater than we think. Turbocharging alone reduces fuel consumption by 20 percent. That can be further reduced using hybrid technology. If you take the 28 million cars produced in China, a reduction of 20 percent of each engine's fuel is equivalent to producing 2.8 million cars with zero fuel consumption and emissions. That would be significant!

Government attitudes are a weather vane for automobile companies. With their instinct for profits, companies will neglect innovation on energy savings and emissions reduction for traditional vehicles when the official bet is on pure electric vehicles, and when huge subsidies are thrown at them. Once a new generation of low-cost, fuel-efficient cars or fuel-powered electric drive vehicles from foreign manufacturers dominate the global market, the technology gap will widen and Chinese cars will lose their hard-won market dominance. That would be a pity.

3. Hybrid power is the way forward

Toyota's 20 year obsession with hybrid power

In my view, hybrid power has proven to be the most mature, feasible, effective, and most influential emissions reducing and energy saving technology.

Toyota began R&D on hybrid vehicles in 1997 and has persisted with it. I call it the Toyota path forward. It has stayed on that path for 20 years, extending it to the core technologies of pure electric, plug-in hybrid, and hydrogen fuel cell vehicles.

I test drove the first-generation Toyota Prius hybrid in Tokyo in 2000. The car had an electric motor, battery, and electronic control device, in addition to a gasoline engine. It used energy recovery technology to capture waste energy from braking and coasting, which was stored in the battery. Acceleration was accomplished solely with the battery or together with the gasoline engine. Energy savings and emissions reduction were achieved through electronic control technology. It was the first time electric drive had been introduced in the history of combustion vehicles, and it was a milestone. Toyota called it hybrid fuel-electric power.

I went to the Frankfurt Motor Show in the autumn of 2007. The global auto industry had gone along with Toyota's Hybrid Technology. Each company promoted one or two hybrid vehicles, but they all lacked technical capability and a commitment to energy saving. It was the fashion to promote smart, pure electric vehicles. But then the global financial turmoil of 2008 suddenly halted the promotion of hybrids.

I went to Toyota again in 2011 and met with Uchiyamada Takeshi, the Executive Vice President (Deputy General Manager) of Toyota Motors, known as the father of hybrid power. I asked him if Toyota was affected by the enthusiasm for electric cars in China, and what impact it would have on Toyota's hybrids. He seemed unmoved by it and said Toyota would keep working on hybrid technology.

Uchiyamada drew me a sketch on a piece of paper. With the hybrid electric vehicle (HEV) as a basis, battery capacity would be expanded and an electric plug would be added, leading to the plug-in hybrid electric vehicle (PHEV). Removing the gasoline engine, retaining the electric motor, battery, electronic control, and increasing the charging function, would result in pure electric vehicles (EVs). Replacing the engine with a hydrogen fuel cell stack, plus two hydrogen storage tanks, would lead to the hydrogen fuel cell vehicle (FCEV).

This drawing, which I have kept, became Toyota's well-known new energy electric drive solution, and was close to what the global automotive industry agreed on.

I was invited back for a private discussion with Uchiyamada on 28 November 2017. He was then President (Chairman of the Board) of Toyota Motor Corporation. He was the first president to be an engineer.

He spoke about the urgent need to solve resources and environmental problems in the 21st century. By choosing hybrid electric power, electrification was combined with the traditional power train without special infrastructure. It reduced fuel consumption and carbon dioxide emissions as soon as customers bought it. That's why Toyota chose hybrid power.

Uchiyamada told me that in the 20 years since 1997, Toyota had sold 10.9 million hybrid vehicles worldwide, reducing carbon dioxide emissions by 85 million tons. He said that when hybrid power was first developed, the goal was only to save energy and reduce emissions. Toyota realized that batteries, motors, and electronic controls were three factors which couldn't be left out. Experience from 20 years of R&D and mass production were the foundation for all electric vehicles. All types of hybrid power are electrically propelled and have energy

Fifteen million Toyota and Lexus hybrid models have been sold worldwide since 1997. The fourth-generation Prius in the picture has a fuel consumption of only 2.68 liters per 100 kilometers.

recovery. This gives Toyota greater confidence in its R&D on new energy electric vehicles.

An indomitable spirit of craftsmanship

The hybrid car represented by the Prius was forged through four generations. Toyota's spirit of traditional craftsmanship has always been praised. I saw the core of Toyota's hybrid technology, its motor, battery, and electronic control technology, at a closed-door technology show. Four generations of progress were on display, visible at a glance.

To reduce the bulk of the motor, the copper wire coil is changed from a circular to a square cross section, and the heat dissipation is also changed from external water cooling to internal oil spraying. Each generation of power semiconductors in the electronic control systems' Power Control Unit (PCU) is overlaid with many pioneering technologies. Finally, high density and low loss are achieved with bulk silicon wafers.

I was impressed by Toyota's concern for consumer cost and experience. Toyota's hybrids, plug-in hybrids and hydrogen fuel cell vehicles mostly use nickel-metal hydride batteries instead of lithium batteries. President Uchiyamada said Toyota has also mastered advanced lithium battery technology, but nickel-metal hydride batteries are low cost, technically reliable, stable with repeated charge and discharge, and more environmentally friendly to manufacture and recycle. Toyota foresees that rare lithium resources will become even scarcer when the number of electric vehicles in the world exceeds one million.

Toyota has come a long way in the past 20 years. It has perfected hybrid technology through 10 million orders. Motor power has increased 200 percent and bulk has been reduced by 50 percent. Battery weight has been reduced by 30–50 percent and size by 60 percent. PCU energy consumption has been reduced by 80 percent and size by 50 percent. It leads in durability, reliability, cost, and overall strength of its electrical components.

Is a hybrid an electric vehicle? According to China's new energy electric vehicle (NEV) classification, only pure electric and hydrogen fuel cell vehicles are considered electric vehicles. Hybrids and plug-in hybrids are excluded. They cannot benefit from any policy preferences or financial subsidies.

That is understandable. In the minds of Chinese people, an electric vehicle is one which gets its power from an external electrical source. Toyota's hybrid energy recovery power generation drive is something else, no matter how good the energy savings and emissions reduction.

Toyota's hybrid technology threshold is too high. It is too conservative about technology transfer and cooperative development. And subsidizing a foreign company seems pointless to the Chinese. It isn't a question of technology, but of policy.

Uchiyamada regretted that hybrids are not being treated as new energy vehicles in China. However, when a consumer buys one, they save energy and reduce emissions as soon as they are on the road. Therefore, he said, Toyota must stay on this path in China even if there are no preferential policies.

Car owners will be pleasantly surprised driving a hybrid vehicle, more so than with a pure electric or an ordinary combustion vehicle. The latest generation of Prius has a fuel consumption of 3.7 liters per 100 kilometers and an electric drive contribution of 40 percent. It is second to none and reaches the lowest 2020 CAFE[1] value set by the Chinese government. Tangible energy conservation and emissions reduction are being made as more Toyota hybrid vehicles enter the market.

The principle of hybrid technology — recovering energy wasted in idling, braking, and coasting — has transformed concepts of automobile propulsion. It will be applied to all attempts at energy saving and emission reduction in the 21st century.

---

[1] Corporate Average Fuel Economy (CAFE) is a concept developed in the US which China has adopted in its regulations as Corporate Average Fuel Consumption (CAFC). Maximum fuel consumption limit values are indexed to vehicle weight and have two sets of limit values, one for regular cars and one for special cars. Compliance will be determined by the CAFC of manufacturers for a given year across their entire product line.

Known as the father of hybrid cars, Uchiyamada Takeshi was the first engineer to serve as Chairman of Toyota. In November 2017, he invited the author to Tokyo to talk about their hybrid approach.

Why plug-in hybrids are becoming more popular

GM's Chevrolet Volt extended range electric vehicle, which was first introduced at the Beijing International Auto Show in 2008, is actually a PHEV. The battery and electric motor are its main motive energy source, and it has a small backup gasoline engine. Its lithium-ion battery propels it up to 60 kilometers on a single charge from an external power source. When the electric power is about to run out, the gasoline engine starts automatically. However, the gasoline engine doesn't drive the wheels, it drives a generator to generate electricity which propels the vehicle. This allows the Volt to be driven more than 450 kilometers.

At present, two main categories of plug-in hybrids are viewed positively in global markets. One is like the Volt, which uses a small gasoline engine to extend its range. The other is like the plug-in Toyota Prius, which has a gasoline engine and a small battery powered from an external source to reduce fuel consumption and extend its range by 50 to 80 kilometers.

Both technologies are practical. Costs can be controlled by limiting the size of the battery. However, officials in China won't subsidize either of them because they don't believe they are a path to pure electric vehicles, and even undermine pure electric vehicles. When subsidies for electric vehicles are reduced in 2020, plug-in hybrids will become popular because they are practical and efficient.

The second-generation Chevrolet Volt extended range electric vehicle can go 53 miles on electric power only and 420 miles overall.

4. Pure electric vehicles make a Great Leap Forward

Overtaking on a curve

There have been three waves of interest in pure electric vehicles in China.

The first was in the 1990s. Chinese science and technology departments liked electric vehicles because Chinese traditional automobile technology was decades behind Europe and the US. Electric vehicle technology hadn't made breakthroughs, so China and the West were at the same starting point and a newcomer could move ahead. That was the basis for the theory of Chinese electric vehicles "overtaking on a curve."

Electric vehicles were included in the key national scientific and technological research projects throughout the Eighth and Ninth Five-Year Plans[2]. I covered the inauguration ceremony for the National Electric Vehicle Demonstration Zone on Nan'ao Island, Shantou in June 1998. Ten domestic electric vehicles, 5 Toyota RAV4s, and 5 GM EV1s were put into operation in the demonstration area. When I went again two years later, the Toyota electric cars had been running for 100,000 kilometers without major problems except for changing their batteries. Major battery explosions occurred 10 times and minor faults occurred repeatedly on the domestic cars, such that drivers were unwilling to drive them. An expert said, "There is no such thing as the same starting line. When it comes to electric vehicles, we are still in elementary school. Foreigners are far ahead."

There was a second wave of interest after the 2008 global financial crisis. Nissan's chief representative in China met with me after the Spring Festival in 2009 and asked me about demonstrating its Leading Environmentally-friendly Affordable Family (LEAF) electric car in several major cities in China. Nissan sponsored a goodwill seminar for senior officials from the Ministry of Science and Technology (MOST), the Ministry of Finance (MOF) and the Ministry of Industry and Information Technology (MIIT) at the Diaoyutai State Guesthouse in April 2009. But a whirlwind of activity on electric vehicles set off by a group of officials and scholars left Nissan out.

Pure electric vehicles are just one of many new energy options in developed countries, but China has singled them out as the only option. The first Automotive Industry Development Plan issued in

---

[2] The Five-Year Plans are a series of social and economic development initiatives issued since 1953 in the People's Republic of China. In March 1991, the fourth session of the Seventh National People's Congress (NPC) approved the State Council's Report entitled "The Ten-year Layout for National Economy and Social Development and 8th Five-Year Plan". Under Deng Xiaoping's leadership, this Plan marked the start of a new phase in China's development. The 9th Five-Year Plan was adopted in 1996 and ran until 2000, and was the first medium-length plan which provided a development strategy for the transition from the 20th century to the 21st.

2009 stipulated that by 2011 the output of electric vehicles would reach 500,000, making up 5 percent of total car sales.

Some experts believed there was a lower threshold for pure electric vehicles than for hybrid and hydrogen energy, and we could do it if we wanted to. There were already millions of low-speed electric vehicles in China they said, and one was the same as another. But a lot of money had been wasted and plans had basically failed because the gap between China and the rest of the world was seriously underestimated. 500,000 pure electric vehicles were planned, but only a few thousand came into the hands of consumers by the end of 2012.

China is wealthy now. Government subsidies for electric vehicles have increased from 10 billion yuan to 100 billion yuan. Multinational companies are developing pure electric vehicles for the Chinese market. I asked a researcher at a multinational company why their attitude towards electric cars had changed 180 degrees. He replied that with the Chinese government investing heavily in the EV market, it was too good an opportunity to miss.

Nearly every multinational company is pushing an electric vehicle. Mercedes-Benz and Audi moved their annual Technology Day to China, focusing on EVs. Even Volkswagen, which has always preferred diesel engines to save energy and reduce emissions, made a high-profile statement: "Electricity helps Volkswagen's heart beat."

Audi invited officials and experts in charge of Chinese electric vehicles to its Ingstadt R&D center in 2010. They were shown innovations in batteries, integrated controls, and motors, as well as R&D methods and verification processes for electric vehicles. They realized that Germany was far ahead of China and there was a lot they didn't know. Overtaking on a curve was just talk.

Despite that, a third wave began in 2013. Electric cars took another leap forward, prompting a global electric car boom.

Major global automotive groups didn't favor pure electric vehicles at first, but after seven or eight years they finally realized what had been staring them in the face. This huge car market was in love with pure electric vehicles.

Manufacturers turned their attention to pure electric vehicles beginning in 2016. For the first time, China was affecting global trends in new energy vehicles.

Batteries are still holding us back

Americans built the world's first electric vehicle as early as 1834, nearly half a century before the advent of the automobile. However, the technology progressed slowly and production of electric vehicles was negligible until 1960.

Fears of oil shortages and an awakening of environmental consciousness in the second half of the 20th century led Toyota, GM, and other multinational companies to invest heavily in electric vehicles. There was a global fever for EVs in the 1990s, and California set a target for sales of 15 percent by 2003. However, that generation of electric vehicles which used nickel-metal hydride batteries was taken off the market in 2003 because of high prices, long charging times, and short battery life.

The low energy storage density of batteries held back electric vehicles for nearly 200 years. Cars caould only run for a few dozen kilometers with hundreds of kilograms of batteries. Even though the energy storage ratio increased significantly over the past decade, from the earliest lead-acid batteries and later nickel-metal hydride batteries to lithium and lithium-ion nickel-manganese-cobalt ternary batteries, material costs also rose correspondingly.

I interviewed Mr. Weber, Mercedes-Benz Group Vice President of Technology, in Stuttgart in the summer of 2008. He said, "The key to a breakthrough is improving battery storage. The energy storage ratio of gasoline to lithium batteries is currently 100:1. The goal is to improve it to 100:10 so that electric vehicles become practical. Mercedes-Benz maintains relationships with battery companies, including China's BYD,[3] to see if there is a substantial breakthrough."

---

[3] BYD Co Ltd is a Chinese manufacturer of automobiles, battery-powered bicycles, buses, forklifts, rechargeable batteries, trucks, etc, with its corporate headquarters in

The capacity of lithium-ion batteries has tripled since 2008, but no substantial breakthrough has been made. The energy storage ratio is far from reaching the 100:10 threshold with petrochemical fuels.

There are reports that the energy storage ratio for lithium sulphur batteries is ten times that of lithium batteries. And it is hundreds of times more with graphene batteries. The future looks bright, but there is a long way to go.

As well as safety risks (a large-capacity phone charging pack can't be taken onto an airplane), there is severe pollution from batteries during manufacturing, and especially when they are disposed of. These will become major problems for us.

Misconceptions about endurance

The subject of range is often raised when electric vehicles are assessed. Officials also take note of range in their evaluations. The cruising range depends on the size of the battery, not technology. A conventional car with 50 liters of petrol in the fuel tank can easily run up to 500 kilometers, while an electric vehicle must carry at least 800 kilograms of batteries for the same range. People often overlook the fact that electric vehicles with higher mileage have lower energy efficiency. It's like carrying a month's rations when you go out for a cup of coffee. High efficiency is only possible with small batteries and a short range.

The product distribution of electric vehicles is shaped like a dumbbell. At one end are toys for the rich — super SUVs that runs at high speed, are luxurious, and for which cost is no object. At the other end are cheap, convenient, mini-cars for ordinary people — efficient for short distances and complementing traditional cars. Development of small and micro-models has been the global focus, and more than 80 percent of them have been of this type. Most of them carry one or two passengers, with a range less than 100 kilometers. Their function as personal, short distance vehicles has become apparent.

Shenzhen. It has two major subsidiaries, BYD Automobile and BYD Electronic. It was founded in February 1995.

On the diagram for new energy, a short line is drawn next to the pure electric vehicle. It shows that it is not a replacement for traditional cars. Buy a small electric car for short, daily trips. The battery is small, charging time is short, and the current doesn't need to be high. Slow off-peak charging is economical and extends battery life.

People are anxious because construction of charging terminals isn't keeping up with the surge in sales, not to mention that charging times are several hours compared with a few minutes to refuel a car with gasoline.

Dirty electricity and the difficulty in fulfilling the zero emissions dream

People often lose sight of the fact that whether or not electricity used as a secondary source is clean depends on whether the primary energy used to generate it is clean.

In an interview with Volvo in Sweden, I was told that for each pure electric vehicle it sells, the user is required to sign an agreement guaranteeing that the source of its electricity is clean. If the user doesn't do it, the company prefers to lose the sale.

It's clear what clean energy is in Europe. It is power generated by wind, tides, hydro, solar, and renewable bio-organisms. Electricity generated from coal is referred to as dirty because coal emits a large amount of harmful substances and carbon dioxide.

In Sweden, hydro-power provides 50 percent of the electricity, 45 percent comes from wind and bio-energy, and only 5 percent from coal. If you operate a pure electric vehicle there, you have a good chance of achieving zero emissions.

In the UK, the last 12 thermal power plants will be closed by 2023, ending 200 years of reliance on coal. £24.5 billion has been invested in nuclear power plants to replace them.

In China, according to State Grid statistics in 2010, 82 percent of electricity was from coal, and it was inferior coal with high sulfur and ash content. Seventy-one percent of the electricity supply will still come from coal by 2020. Less than 30 percent will come from hydro, wind, nuclear, and solar power.

Take the domestically produced 2009 Golf 1.4TSI with its advanced technology engine as an example. For the entire cycle of energy consumption, carbon dioxide emissions are 143 grams per kilometer. With an electric vehicle of the same power, emissions are 171 grams, and sulfur dioxide emissions are greater.

Globally, it's worth hoping that the zero emissions of pure electric vehicles will come about. But in China the incontrovertible fact is that there are only two energy sources for automobiles: gasoline and coal. By relying on coal to generate electricity, energy conservation and emissions reduction is effectively non-existent.

One piece of good news is that the four major power plants that supplied most of the electricity to the capital ceased coal-fired power generation in 2017, and four major gas-fired thermal power plants were completed. That might not make electric vehicles emissions free, but at least sulfur dioxide will drop significantly.

We have to ask, when annual output of electric vehicles grows from hundreds of thousands to half the total output of 30 million, will China's electric power industry be up to it?

Fiscal subsidies and administrative intervention haven't proven effective

Electric vehicles surged to another high at the beginning of 2013, meeting the national strategic goal.

The government set a clear goal of 500,000 new energy vehicles (pure electrics and plug-in hybrids) by 2015, with annual production capacity reaching 2 million, and sales exceeding 5 million by 2020.

To achieve these goals, government departments served up two substantial inducements. The first was financial subsidies. The Central Government provided a subsidy of 35,000 to 60,000 yuan for each pure electric vehicle according to its range, and local governments provided roughly the same. The subsidies were as much as one million yuan for pure electric buses. The second was administrative intervention. In some big cities where car purchases were restricted, the issuance of traditional car licenses was further restricted. Meanwhile, electric vehicles were unrestricted and administratively fast-tracked. This led consumers to make the switch.

This two-pronged promotion of an industrial product was unprecedented. Traditional automobile factories had to build electric vehicles to stay in the game and invested huge amounts of capital in the effort. Electric vehicle sales exceeded 500,000 for the first time in 2016. And China became the world's leading producer.

How good were these electric vehicles technically? Few of my friends who bought them were smiling. Most domestic electric cars were still low-end products. Their technology and quality were not up to the mark and they were not nice to drive. Because my friends were worried about the range, they didn't turn on the air conditioner or the heater. And the cars couldn't be started when it's too cold.

As of 2 March 2017, the NDRC issued production certificates for 11 new electric vehicle companies. There were another 200 new electric car companies waiting in line across the country. If the various electric vehicle projects of traditional automobile companies were counted, total production capacity would exceed 10 million vehicles and investment would exceed 300 billion yuan. There was a lot of worry that the chaos of overheated investment and overcapacity would repeat itself in the field of electric vehicles.

Multinational companies each had their own R&D focus, as well as a full complement of core technologies for electric vehicles, including batteries, motors, electronic controls, and lightweight vehicles. Mercedes-Benz introduced its EQ series electric car, BMW had its i-series, Volkswagen had its I.D. series, and Tesla had its Model 3. It would be a waste of effort if we created a market, only to be outdone by better products.

Domestic electric vehicles began to show some promise in 2018. Specialist EV brands such as BAIC New Energy, BYD, and ORA[4], increased production, as had some companies using traditional vehicles as platforms. NIO, Xiaopeng (Xpeng or XMotors.ai) and other newly

---

[4] ORA is a sub brand of Great Wall Motors specializing in electric vehicles.

AIWAYS, meaning AIONTHEWAY and founded in 2017, is one of many new EV powerhouses in China. AIWAYS is speeding up the evolution of the EV industry with smart manufacturing, products, and services.

influential enterprises were competitive with their use of capital and smart connected technology. I wish them well.

5. Hydrogen energy is the ultimate solution

Hydrogen redefines the car

Hydrogen is an inexhaustible source of energy, and only emits air and water. It is virtually emissions free. Therefore, the global automotive industry regards hydrogen fuel vehicles as the ultimate solution.

Hydrogen fuel cell electric vehicles are abbreviated as FCEV. Many people just call them FC (fuel cell) vehicles, but it specifically means a fuel cell vehicle with hydrogen energy.

FCEVs have become the pinnacle of technology for the world's most powerful automobile manufacturers.

I examined a Mercedes-Benz first-generation FCEV at its Stuttgart R&D Center in July 1997. Fuel cells were still large then. Except for the driver's seat, the entire cabin of a vehicle about the size of an MB100 van was completely filled with a huge fuel cell stack. This was probably the first hydrogen fuel car a Chinese person had ever seen.

I also test-drove GM's third-generation hydrogen fuel cell vehicle, the HydroGen 1, which was demonstrated to the Chinese automotive industry and S&T community at the Ministry of Communications test site in Beijing in October 2000. It was based on Opel's Zafira, complete with steering wheel, brakes, clutch, and dashboard. When they began to explain the principles of hydrogen fuel cells, Chen Zutao, one of the senior people there, dragged me to the front row of the venue.

A fuel cell is composed of an extremely thin electrolytic membrane sandwiched between electrical plates. Platinum layers are coated on both sides of the membrane to form the positive and negative poles of the battery. Hydrogen gas is broken down into electrons and protons by a platinum catalyst. The protons can pass through the electrolytic membrane, and water vapor generated from oxygen in the air is discharged. The electrons blocked by the electrolytic membrane are collected to generate electricity. The stored current drives the motor to propel the vehicle.

My most memorable hydrogen energy experience was driving GM's Hy-wire, its second-generation hydrogen-powered concept car, at Milford Proving Grounds near Detroit in July 2003. The Hy-wire is very different from a traditional car. It not only uses hydrogen power, but also forsakes traditional components such as the steering wheel, gearbox, and drive shaft. Control is transmitted by wire as in computers, completely changing the way we drive. Hy-wire with its hydrogen power plus drive-by-wire truly redefines the automobile.

The shiny Hy-wire is the size of a streamlined luxury car. There is no B-pillar[5] to support the roof and the front and rear doors open opposite

---

[5] The B-pillar is the post between the front and back seat doors of a sedan, connecting the roof to the body.

each other. Hy-wire's hydrogen fuel cells, electric motor, and drive-by-wire system are concealed under the floor. Above the floor is all seating space. When you sit in the car, the front window glass extends all the way to your feet. It has the spacious and transparent feeling of a loft. Hy-wire doesn't have a steering wheel, only a double-handed lever like a game console. Hold the lever and gently move it forward to start and accelerate; pullback to decelerate and brake. It turns easily when the lever is moved left or right. The lever can be moved between the left and right front seats, to switch drivers. A video screen in the middle of the double-grip lever is the rear view mirror. With no accelerator and brake pedals, your feet are completely free.

The test track is circular, and the curves are marked with red traffic cones. I deliberately approached a cone to test the accuracy of the controls. It was really a pleasure to drive. Hy-wire makes almost no noise, and you can barely hear the light hum of the motor. It moves smoothly like a canoe on the water.

GM launched the Chevrolet Sequel, its third-generation FCEV, two years later. It's more like an ordinary SUV in comparison to the Hy-wire concept car. It still uses drive-by-wire, with braking, operating, and suspension all changed to electronic control. It travels 480 kilometers on a single fueling. It has only one-tenth of the moving parts of traditional cars.

As concept cars, FCEVs are expensive. GM's vice president of R&D announced in 2003 that its goals for FCEVs were mass production, a reasonable price, and profitability. It wanted to be the first company to sell 1 million of them by 2010. President Bush was interested in hydrogen-powered vehicles, and large numbers of hydrogen refueling stations were planned.

The mainstream is still traditional automobiles and although there are hydrogen-powered vehicles, R&D investment is huge, and support couldn't be given to a single manufacturer. The financial storm of 2008 broke GM's capital chain, it went bankrupt, and had to reorganize. Newly elected President Obama's interest turned to electric vehicles. Hydrogen-powered vehicles failed just as they were on the verge of success.

Hy-wire redefines the car with hydrogen power and drive-by-wire. Drive system is under the floor, with full front vision.

The game console-like control can be moved from left to right.

Without a B-pillar, front and rear doors open opposite each other.

China can boast when hydrogen energy regains momentum

The good news is that in the past decade, Mercedes-Benz and Volkswagen in Germany, Toyota in Japan, and GM in the US have started a new round of R&D and marketing of hydrogen fuel cell vehicles.

Mercedes-Benz launched the F-CELL, its next-generation hydrogen fuel cell vehicle, in 2009. It is touted as the world's first mass-produced hydrogen fuel cell vehicle. During the 2011 Shanghai International Auto Show, a demonstration F-CELL arrived in Shanghai after a 125 day trip across four continents. I was thrilled to test drive it in the Expo area. It was quiet, smooth, and fast. It used the latest hydrogen fuel cell drive system with an output of 100 kilowatts and a range of 400 kilometers, and it took only 3 minutes to refuel.

I test drove the MIRAI, Toyota's latest mass-produced FCEV, at its R&D Center in Changshu in October 2017, and visited the hydrogen refueling station there.

The MIRAI FCEV looked like a Toyota Prius hybrid inside and outside. It was like driving a high-end pure electric vehicle. There was no sound and it accelerated particularly well. A refill only took three minutes. The 500 kilometer range was as good as a traditional car.

I took a closer look at the structure of the MIRAI's hydrogen fuel cell power train. There are two hydrogen storage tanks at the rear of the car. Under the front seat is what corresponds to the core component of a traditional automobile engine — the fuel cell stack. Hydrogen enters its 370 layers and combines with oxygen to generate electricity and water. The boost converter at the front of the car raises the output of the battery stack to 650 volts. The current enters the AC synchronous motor located in the front compartment, propelling the car. The maximum power of the motor is 113 kilowatts, and the maximum torque is 335 Nm. Next to the motor is a power control unit (PCU), an electric control device which is used to control the power output and discharge of the fuel cell. Using the principle of hybrid power, there is a nickel-metal hydride battery at the rear of the car to recover energy that might be wasted during braking and coasting. It also generates power for the electrical motor.

Nakao, General Manager of the R&D center, said that Toyota began to develop FCEVs in 1992, and they were launched commercially in 2008. Costs were high and they were expensive. With technological advances, the cost has been reduced to one twentieth of what it was. MIRAI, released in 2015, sells for 7.25 million yen in Japan with tax. That's about RMB 420,000. Three thousand of the vehicles were sold in Japan, Europe, and the US in 2017.

Two other factors are the production of hydrogen and the construction of refueling stations.

There is as much hydrogen on earth as we need. Middle school students know that water molecules are composed of two hydrogen atoms and one oxygen atom, so electrolyzing water is an important way of producing hydrogen. The electricity used to electrolyze water can come from coal, or from clean wind or solar energy. Hydrogen is a by-product of the steel and chemical industries. It can be extracted from low-quality lignite, and from bio-gas.

Although the technical threshold for hydrogen fuel cell vehicles is high, China's R&D started early enough, and it still has a role to play. The School of Automotive Studies at Tongji University began to develop the Chaoyue-1 hydrogen fuel cell vehicle and took an independent technical route in 2006. In the 2007 Michelin Bibendum New Energy Vehicle Challenge[6], Tongji Chaoyue-3 succeeded in having the lowest hydrogen consumption. Professor Yu Zhuoping, Dean of the School of Automotive Studies, told me that hydrogen sources can be diversified. The hydrogen produced as a by-product in the steel and chemical industries in Shanghai alone can support 400,000 vehicles. And hydrogen is even cheaper than gasoline.

Although China has been promoting pure electric vehicles, it is convinced that FCEVs are the ultimate solution for energy conservation and emissions reduction. Great Wall Motors, a rising star among Chinese brands, has begun to systematically position itself on new energy, and will launch new vehicles in 2022. According to the hydrogen fuel cell vehicle development plan released by the Ministry of Industry and Information Technology in October 2016, the number of fuel cell vehicles will reach 5,000 in 2020, 50,000 in 2025, and 1 million in 2030. In the same period, there will be 100, 350, and 1,000 hydrogen refueling stations, respectively. That will be worth boasting about if we can achieve it.

07 Farewell, Mr. Wagoner: An impression of Rick Wagoner

I went to Sina.com[7] to do a live video broadcast of the Shanghai Auto Show on 30 March 2009. Before entering the studio, Su Yunong,

---

[6] The Michelin Challenge Bibendum or Movin'On (since 2017) is a major annual sustainable mobility event, sponsored by the French tire company Michelin. "Bibendum" is the name used in France for the iconic figure known in English language countries simply as "The Michelin Man".

[7] SINA is the most recognized internet brand name among Chinese communities globally. SINA's portal network consists of four destination websites dedicated to the

Editor-in-Chief of Sina Auto[8], told me, "I can't be with you now. I just heard that GM's President Wagoner has resigned and I have to put together a report on it immediately."

The Rick Wagoner I knew

I felt sad the day I heard Wagoner had finally resigned. President Obama had stated that the bosses of the three US car companies should step aside if they couldn't make changes. I felt there were complex reasons on how the US Big Three got to where they were, but it was too much to expect immediate results from reorganizing and changing leaders in a crisis. However, US$17.4 billion in financial help had been squandered, and GM and Chrysler hadn't come up with a decent solution by the 1 April deadline. Instead, GM asked for another US$16.6

Chinese communities across the globe: Mainland China (www.sina.com.cn), Taiwan (www.sina.com.tw), Hong Kong (www.sina.com.hk), and overseas Chinese in North America (www.sina.com). Each destination site consists of Chinese-language news and content organized into interest-based channels.

[8] SINA Auto offers the latest automobile-related news and service information to auto shoppers and car enthusiasts, including current information on pricing, reviews, and featured guides.

billion to turn itself around. But the new President hesitated to put it to Congress. Government officials called for Wagoner's resignation on 27 March. And, on the 30th, the man who had insisted he was the right one to lead GM out of its predicament had to resign.

I've met many global auto industry leaders in 30 years of interviews. Wagoner was one I could call a friend.

I went to the GM Global Forum in Italy in 2000. At the terraced venue, Wagoner, who had just become CEO, was wearing a light blue shirt and khaki trousers. While presiding over the meeting, he ran up and down, passing the microphone to the speakers. I had been to GM global conferences almost every year since then, discussing topics like design, hydrogen energy, and product strategy. I also had numerous interviews and exchanges with Wagoner. What most impressed me was that the CEO of the world's largest company didn't put on airs. He waited in the line with his plate in his hands at the buffet table. He gave up his seat to the media and stood at the back listening to the lecture. That couldn't happen in a Chinese corporate context.

I went to GM's Global Product Symposium in Santa Barbara, California in August 2002. At a seaside garden as darkness fell one evening, our cameras were taken away and GM displayed 15 concept vehicles and new cars for the next year.

I became interested in a small multipurpose vehicle with a fiberglass body because I had heard that this new car, code-named AFC, had been developed for Chinese and Asian families. It emphasized practicality and low cost, but was not particularly elegant. While I was looking at it, Wagoner came over and asked me what I thought? I softened my answer, saying it was good that GM had developed this completely new small concept car for a larger consumer group. But, if such a model was produced in China, there would be a dilemma between controlling costs and satisfying Chinese sensibilities. Frankly speaking, building low-cost cars in China was not GM's long suit.

I said that private cars were just taking off in big cities, and car buyers were still in the top half of the consumption pyramid. I suggested that GM should first bring its top level flagship products to China, such as the Cadillac. Its latest model, the CTS, was a great hit and should represent the GM brand. Wagoner listened intently. He probably hadn't expected me to say that.

I thought it was just a few casual comments, and that would be it. However, at the closing ceremony, Wagoner said friends in the media had given him some new ideas.

There was soon proof that bringing the Cadillac into China had seriously begun.

Wagoner officially announced to the media in Beijing in November 2003 that GM would produce the Cadillac in Shanghai. He was careful to add, "Some of you suggested this. That's why we decided to do it."

Whenever Wagoner met me after that, the first thing he said was, "Have you got any new suggestions, Mr. Li?"

He did his best for ten years

It's unfair to say GM hadn't focused on developing new energy vehicles. The first electric vehicle the Chinese ever saw was the GM Impact which debuted in Beijing in the 1990s. Later, five GM EV1s were put through long-term testing at Nan'ao Island National Electric Vehicle Demonstration Zone in Shantou. GM's investment in hydrogen-powered vehicles, the ultimate solution for energy savings and emissions reduction, was the largest in the global automotive industry. It went the furthest and was the most successful with hydrogen fuel cell technology. The original gearbox and transmission system were replaced by electronic modules. When Japanese hybrid power emerged in the early 2000s, Wagoner changed his dismissive attitude towards it and began using it in high fuel consuming vehicles like pickups and SUVs. GM has always redefined automobiles.

Wagoner came to Beijing before the 2008 Beijing Auto Show to show me the Volt, GM's extended range electric vehicle. I thought it was just a new hybrid car, but he said to me, "GM's new energy technology is not limited to internal combustion engine optimization, hybrid power, and hydrogen fuel cells. The Volt is a step we've before fuel cells are commercialized. It's different from a hybrid that relies on fuel. ItI's a plug-in electric vehicle, and the internal combustion engine is only supplementary. It gives electric vehicles a way forward by making up for the shortcomings they have had."

Wagoner's and GM's thinking on new energy systems seemed feasible. Unfortunately, heroes are often judged by their failures as well as their successes. Wagoner's new ideas were ignored in the headlong rush to pure electric vehicles.

In his ten years, Wagoner reorganized GM's brands and globalized its products. Through the development of a modular architecture, its brands provided models that met consumer needs in different markets. GM also successfully entered the field of small cars where it previously had no presence.

As the 21st century began, Wagoner and former Chairman Jack Smith were pressured to clinch a joint venture in Shanghai to build GM's most modern factory. Not only did it introduce Buick's latest models, it also established China's first joint venture car research and development center. Shanghai GM and SAIC-GM-Wuling[9] have become China's most successful car companies. Some of the credit for that undoubtedly goes to Wagoner.

A bad bet on the capital market

When I saw Wagoner at home and abroad in those years, he looked increasingly stressed, and his smile was less relaxed. Cao Xueqin said it best in *A Dream of Red Mansions*: "The great suffer great hardships." GM, which had been at the top of the global automotive industry for more than fifty years, seemed shaky. Huge benefits paid to retired employees were unsustainable. This was compounded by the powerful United Auto Workers' (UAW) continued fight for higher wages and benefits, which they won time and again. That made the labor costs of the Big Three US automobile companies, including GM, 70 percent

---

[9] SAIC-GM-Wuling Automobile is a joint venture between SAIC Motor, General Motors, and Liuzhou Wuling Motors Co Ltd. Based in Liuzhou, Guangxi Zhuang Autonomous Region, in southwestern China, it makes commercial and consumer vehicles sold in China under the Wuling and Baojun marques, respectively.

higher than their Japanese rivals. Coupled with the huge development costs of new energy and new technologies, finances were increasingly stretched. Wagoner, who had previously been CFO, had to bet on Wall Street, hoping to play the "concept" card to get investment from capital markets and sustain his cash flow. However, the unexpected financial tsunami swept everything away, bringing all his efforts to nothing. After first receiving government financial help, Wagoner also counted on creditors agreeing to swap debt for equity on two-thirds of the US$27 billion debt, but neither the creditors nor the unions would go for it. The end had come, and there was no hope of recovery. Sadly, Wagoner had to call it a day.

The day Wagoner resigned, Obama said at the White House that the automobile industry was a pillar of the economy, and the US must not let it disappear. He also warned that if GM and Chrysler couldn't formulate a more effective restructuring plan, the US government might choose to let them seek bankruptcy protection. People in the media asked me how likely GM was to go bankrupt? I said, it was impossible to predict, but several of GM's major brands should be able to survive by restructuring. However, because it operates independently, Shanghai GM, its joint venture in China with quality brands such as Cadillac, Buick, and Chevrolet, would not be affected much.

The impetuous Chinese auto industry has a lot to learn

Wagoner's resignation allowed the US government to give GM more help. But at a deeper level, which single individual can be held responsible for the US automobile fiasco?

For more than a century, the global automotive industry had been shifting, from Europe, the home of automobiles, to the US, Japan, South Korea, and then China. It has been a century-long shift. In the early 20th century, Ford's assembly line production reduced costs and created explosive growth in the US automobile industry. From the 1950s to the 1980s, Japan took the lead with lean production, lower costs, and better quality. Seven or eight years ago, who could have believed that China would become one of the world's greatest auto powers?

China's auto industry was successful by being in the right place at the right time. But its current reliance on the advantage of low labor costs can't be ignored. Brands, R&D, and technology are only just beginning. China's auto industry hasn't fully emerged from a disorganized pattern. Industrial reorganization has a long way to go. There is no place for complacency and schadenfreude. The impetuous Chinese auto industry can learn a lot from GM's mistakes.

The world is getting smaller and I hope Wagoner will be able to come to China often.

Chapter 12

When Will the Dream of Being an Automotive Power Come True?

ccording to Chinese tradition, time is counted in 60-year cycles. Just one 60-year cycle has passed since the first Red Flag sedan, equipped with a V8 engine, was born at FAW in 1958. Those 60 years fall into two parts. For the first 20 years, China's auto industry was in a start-up stage of closed borders and self-reliance. The last 40 years has been a period of rapid development characterized by Reform. Globalization and commercialization have made China the world's major automobile producer, especially since it joined the WTO 20 years ago.

The Chinese auto industry is now at the beginning of its second 60-year cycle, and it is much larger than it was 60 years ago. The problems being faced and levels of support involved are also much greater.

1. Take a deck of cards along when you drive

It's hard to find a panacea for traffic jams

China gave policy support to automobile consumption in response to the global financial crisis. Automobile production weathered the crisis to achieve a substantial increase of more than 30 percent for two consecutive years. Output reached 18.26 million vehicles in 2010, surpassing the US for the first time. In subsequent years, China became the world's largest automobile producer.

Friends often joked that I had opened a Pandora's box of traffic jams by calling for family car ownership. But whether you are white, black, or Chinese, life is better with a car than without one. Now that Chinese people have cars, they don't want to go back to living in a bicycle kingdom.

Whether people accept it or not, China has become an automobile society, like all developed and most developing countries in the world. While enjoying the culture, wealth, and economic and technological development brought by the automobile, we also have to face up to the negative challenges of traffic congestion, emissions, and energy consumption.

There is no panacea for traffic jams in mega-cities like Beijing. The wrong medicine only makes them worse.

One of the earliest major traffic jams I reported on in Beijing happened in 1983, when cars returned from outside the city after National Day. Cars queuing to cross intersections waited for ten or more traffic light changes. Bus speeds fell to less than 5 kilometers an hour. Beijing's motor vehicle ownership was only 250,000 then. This showed the number of cars was not the only factor in traffic congestion.

Beijing started a ten-year road construction campaign in 1984, and a race began between roads and numbers of cars. The 33 km Second Ring Road, the 48 km Third Ring Road, and a hundred overpasses changed the face of the ancient city. Even home-bound old Beijingers went out beyond the Second Ring. The unclogged roads stimulated an increase in vehicles. In the next ten years, roads increased by 3.2 percent a year, but motor vehicles increased by 15 percent. Less than a year after the Third Ring Road was completed, traffic jams at the exits were causing problems for drivers and the Traffic Management Department.

In December 1995, when there were 800,000 motor vehicles in Beijing, consideration was given to levying tens of thousands of yuan in urban expansion fees on newly purchased cars. When the news leaked out, Beijing people queued up to buy cars and get licenses to beat the levy, kicking off an increase in family car ownership. The levy was increased, then readjusted by the State Planning Commission and Ministry of Machinery, and finally shelved.

Provisions on driving cars with odd and even license plate numbers within the Third Ring Road came out after Spring Festival in 1996. They only limited mini cars, jeeps, and station wagons with a displacement of less than one liter. An official came forward to say that Beijing had a special status as the country's political and cultural center, thus services to party, government, and military leaders had to be maintained and these people mustn't be hindered in carrying out their activities. Therefore, the provisions should only apply to small cars. With a bit of black humor, the official said that a driving limit of two and a half days a week would make people more efficient. As it would only apply to private cars, the idea was poorly received and it died quietly two months later.

Two major events brought about extraordinary developments both for transport and private cars in Beijing in 2001. The successful bid one year before for the 2008 Olympic Games allowed Beijing to start large-scale urban infrastructure construction, marked by road and rail transit. And, the National People's Congress passed the Tenth Five-Year Plan that spring, encouraging family car ownership and affirming legislatively for the first time the right of ordinary Chinese people to own cars.

A boom occurred in Beijing and other major cities between 2001 and 2008, as ordinary people bought private cars, and public vehicles were upgraded. These were the Eight Golden Years of traffic management in Beijing. Motor vehicle numbers surpassed two million and then three million in that period, but Beijing resisted the temptation of license plate auctions, in the end maintaining a fragile balance between cars and roads.

When the Olympic Games were held, odd and even number plate restrictions were temporarily implemented within the Fifth Ring Road, and vehicle owners showed understanding and support. Traffic flowed more freely than it had for many years.

The authorities mistakenly thought the reduced congestion brought about by traffic restrictions in 2009 would become the norm. Five-day restrictions were implemented. However, these created the illusion that the roads were now clear. Even more people bought cars, bringing Beijing's car ownership to more than 4 million.

On a drizzly weekend evening on 17 September 2010, the Beijing Traffic Management Bureau's monitor screens flashed red, as congestion paralyzed the city for 9 hours. Seen from space, Beijing must have looked like an oversize parking lot.

On 23 December 2010, Beijing announced that from January 2011, number plates would be allocated by lottery. Lotteries for 20,000 license plates were held once a month. The total number of small passenger cars in Beijing was limited to 240,000 in 2011. In subsequent years, there were fewer cars in the lottery, while more people took part. It was harder to win than a regular lottery.

It is foreseeable that traffic congestion will worsen when authorities proclaim their support for eliminating traffic congestion while they continue to approve the building of skyscrapers in the most congested CBD areas; when people leave home early and return late because they live outside the city but their workplaces are in the city center; and when China's ancient checkerboard road layout has been replaced by a bullseye of concentric rings. It's no wonder we have to take a deck of cards to while away the time in traffic jams.

Officials and scholars talk about Green Travel Consciousness. The share of public transport is more than 60 percent in major European and US cities. It's less than 30 percent in Beijing, despite strong investment in recent years. I doubt that odd and even number plate restrictions will halve the number of private cars, or if public transport can bear twice as much pressure. A white-collar friend responded to the call and gave up her private car to take the subway. It took 40 minutes in the morning and she only just squeezed onto the platform. She said, "Being so crowded is undignified. What sort of green is that!"

When American billionaire Michael Bloomberg became Mayor of New York City, he took the subway to work every day. I have suggested many times that officials in charge of planning and urban transport should experience squeezing onto the subway during rush hour once a week to see what ordinary people feel like. What we want is dignified, green, comfortable, and accessible public transport.

Smart vehicles and ride sharing bring new hope

The driver rotates the seat 180 degrees, leaving the steering wheel behind him, and the car drives itself obediently out of the parking space. It goes forwards, changes lanes, turns, and automatically avoids pedestrians, according to the destination the driver has entered. There are large touch screens on three sides where passengers can communicate with the car, the internet, and the world outside the car, and even replace the street scene outside the windows with images of natural scenery. The driver gives me an authorization and I swipe an icon to go faster or slower.

It's not science fiction. It really happened on my test ride in the Mercedes F015 in San Francisco in spring 2015. Since then, the car has caused a sensation at several international auto shows. It even outdoes futuristic cars in Hollywood science fiction movies.

With continuing R&D, the F015 not only foreshadows smart driving technology after 2030, it also shows what luxury brands will be like in future. The car is now purely electric, but in 15 years' time it may use cleaner energy, such as hydrogen power.

Smart cars are a hot technology worldwide. There are two schools as I see it. One in Europe, led by traditional automobile companies, and one in the US, led by the IT industry. The former comes out of 20 years of innovation in safety and ease of driving, and the latter is based on innovation and the impulse to replace traditional cars. In China, the force of this new momentum is powered by huge amounts of capital.

Smart interconnections are solving traffic congestion and safety problems through big data and cloud computing. I had a smart assisted driving experience when I tested a Tesla Model S on a busy main road in Yizhuang, when I was back in Beijing. I activated the automatic driving system at the bottom left of the steering wheel and put my hands on my knees. The car automatically merged, changed lanes, overtook and gave way. With this system, as with an aircraft autopilot, the driver can free his hands, but can't put on a blindfold and go to sleep. Data shows that the accident rate of Tesla cars in self-driving mode is reduced by about 40 percent. Intelligent cars are divided from low to high from L1 to L5. The Tesla is L3. The Mercedes-Benz F015 mentioned earlier is L5.

Major auto companies in China proclaim that they are no longer car manufacturers, but mobility service providers.

Mobility services integrate traditional cars with mobile internet. Startups, car companies, and investment companies are coming at this from different directions.

Internet-based ride-hailing with drivers' own cars has gained a foothold with a new operating model. Time-share leasing and shared travel have been quietly infiltrating people's lives. There were already 40 time-share operators in China in 2017, with more than 40,000 vehicles in operation, 95 percent of which were electric vehicles.

Will autonomous cars and car sharing be a disaster for car sales? People are no longer eager to own a car. Will car buying behavior be replaced by a car app on a mobile phone? Will car companies be reduced to business-to-business (B2B) companies providing vehicles for ride-hailing?

I interviewed Tesla Global President Jon McNeill in the summer of 2016. He said, "When I came to Beijing, I saw streets full of yellow and orange bikes. In Tesla's view, the sharing economy is an opportunity, and we are only at the beginning."

He also said, "When fully autonomous driving (L5) gets regulatory approval, it means you will be able to call your Tesla from almost anywhere. When you get in the car, you can sleep, read, or do anything else on your way to your destination. You can also easily set a button on your Tesla App to add your car to the Tesla Shared Fleet, so that your car can earn income for you while you are at work or on vacation. That reduces the cost of ownership, and everyone can share a Tesla."

I think it's quite plausible for private ownership and sharing to coexist.

Minister of Science and Technology Wan Gang said that the 5 million cars on Beijing streets caused congestion and took up space when

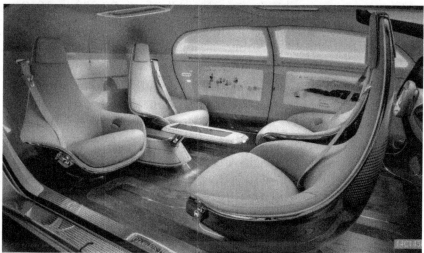

At the 2015 Consumer Electronic Show (CES) in Las Vegas, Mercedes-Benz launched the F015 concept car, which anticipates intelligent driving technology after 2030 and explores the future of luxury.

they were parked. One million autonomous cars would be enough to replace them.

If that happens, traffic jams will become a thing of the past.

2. The late arrival of automobile culture

An industry under strong external constraints

There are strong external constraints on the automobile industry.

China became the world's largest automobile producer in 2011. It is also the world's biggest greenhouse gas emitter and energy user, and the most polluted country on the planet. The next 60 years will bring new pressures.

Energy, the environment, and transport are major bottlenecks for sustainable development of automobiles. The only way forward is to coordinate.

China's car ownership increased rapidly, from 50,000 in 1949 to 200 million in March 2017, an increase by more than 4,000 times. Beijing, Chengdu, Chongqing, Shanghai, Suzhou, and Shenzhen each have more than 3 million vehicles. According to Wang Xia, Chairman of the China Council for the Promotion of International Trade (CCPIT) Automotive Branch, China's car ownership is set to surpass that of the US in 2019. The US has the most cars, with 253 million.

About one-fifth of cities have serious air pollution, and more than one-third of the 113 key cities have air quality that does not meet national standards. Vehicle emissions, coal smoke, and dust are the main sources of air pollution in large and medium-sized cities. People are starting to be critical of the negative effects of the proliferation of cars.

This has put pressure on the large quantities of gasoline consumed and on protecting the environment. However, that pressure is a huge driving force for R&D and innovation. Fuel consumption has been brought down by 50 percent globally in the past 20 years. Emissions and pollution have been reduced by 98 percent. Development of new energy vehicles continues apace. Research in Europe and the US can already be put into practice to mitigate harmful substances with effective air purification devices, making automobile emissions cleaner than the surrounding air.

Having obtained the right to own cars, Chinese people tried to replicate in 20 years a mature car society it took foreigners more than 100 years to achieve. That process has been progressive, difficult, and irreversible. But it is a two edged sword.

Consumption in China is like the Yellow River, shallow when dry, or flooding its banks. No doubt China's move to an automobile society should be a gradual process, but the cars pouring into big cities are not getting the support they need.

Complaints and countermeasures from various sectors, professions, and interest groups in the past 20 years have been impulsive, extreme, irrational, and stopgap. What Chinese people face in the next 60 years is not an automobile industry, but an automobile society, which they will all be a part of.

China's growth has brought with it serious economic and social problems, compounded by a demographic dividend, and costs for the environment and resources. Energy shortages, traffic jams, and environmental pollution are too much for people to bear. Economic transformation is our nation's guiding principle. An overall plan is needed to scientifically forecast and position the development of the automotive industry in the next ten years, starting from overall energy structures, urban planning, and environmental controls.

Developing cars will be like breaking through the sound barrier as car ownership grows in the second 60-year cycle. In a Tsinghua University study conducted in 2010, the limit of car ownership in China was 150 million. That was too conservative and has been far exceeded. According to my prediction, the limit is 220 million, and it will enter a stage of steady development after that. A turning point will possibly be reached in 2018 when growth will slow down.

Promoting coordinated social development and growth in consumption of automobiles will be important issues facing the industry.

The Tour de France from a helicopter. Spectators drive to the track.

Seasonal workers stop for a cup of coffee. They drive to the vineyards of Alsace from all over.

Young people cycling on the streets of Denmark.

Living with a car

Life with a car is certainly no longer the same as life without one.

Cars have flooded into Chinese people's lives in the last 20 years. A country becomes an automobile society when it has 10 to 20 cars per 100 households. This number is now more than 20 in major cities and developed coastal areas, the largest being Dongguan, which passed 60 in 2010. This has brought China to the threshold of an automobile society.

A car is not just a product, but also a way of life. However, the tide has come so fast that many scholars and officials still don't understand what a car is.

Many Chinese people still see cars as industrial products, not realizing that their impact on society is much greater than the product itself. Automobile socialization has increased the scope of people's activities hundreds of times, and it has changed traditional concepts of time and space. It can't be denied that more than a century of global urbanization has been the product of automobiles. The wide range of industries covered by automobiles, the wide application platform they provide for new technologies, and their profound impact on human civilization make it difficult for any other industry to surpass them.

Cars began in Europe more than 130 years ago, grew up in the US, were revitalized in Japan and South Korea, and triggered an economic miracle in China. The global industrial focus shifted eastward, and cars became a complete way of life. The automobile stood prominently on the human landscape of the 20th century. No other invention had such a profound impact on humanity. No aviation, aerospace, high-speed rail, or home appliance societies appeared in the 20th century, although an internet society has emerged in the 21st century.

China is excited about leading the world with pure electric cars. Pollution-free hydrogen energy vehicles are being perfected. Shanghai GM General Manager Ding Lei even talked to me about nuclear-powered vehicles. A few milligrams of nuclear fuel would be enough for the entire useful life of a car.

A photo of Haizhu Bridge in Guangzhou during peak rush hour in the early 1980s stays in my mind. The whole bridge was jammed with bicycles, with a bus like a boat caught in the current.

Forty years later, the bicycles have been replaced by cars. Large cities in China are like most international ones. However, just having millions of cars does not make an automobile society.

Automobile societies were created in Europe and the US over four or five generations, taking more than 100 years. Chinese people can't achieve that in 20 years.

Chinese people like driving big cars and localizing introduced models. Foreigners have finally understood the Chinese characteristic of lengthening everything from small cars to luxury ones.

Chinese people have to put up with traffic jams because they drive competitively and never give way to each other. For several years now, it is newsworthy if the Beijing-Tibet Highway through the Badaling Great Wall Scenic Area isn't blocked for tens of kilometers on public holidays.

Chinese people like to drive copies of well-known international cars, for less than half the price. And there is a range of services to re-badge them. Foreign companies will never win a trademark infringement lawsuit in China.

China has the most road monitoring detectors in the world, and is one of the few countries to have special traffic police, but traffic management is poor. When traffic slows to walking speed on Beijing's Fifth Ring Road, you can be sure that the traffic police must have set up a roadblock ahead to stop trucks from outside Beijing driving on the road illegally.

The Chinese might have learned to love big fast cars from Hollywood movies, but I would like Beijing's urban transport to be modeled on that of Paris. Its main strength is a convenient, accessible subway. Even people with money buy small or very small cars. The streets in central Paris are not much wider than Beijing's hutong alleyways, and they are mostly one-way. Residents park their cars bumper to bumper. With hardly a few inches between them, you have to inch backwards and forwards to get out. It's not a good idea to buy a big luxury car in Paris.

Standing high up on Montmartre, looking at the lines of cars of different styles and colors parked along the roads around the hillside, I wondered which car Amelie got out of in the movie. The combination of cars and Paris made a harmonious and charming picture.

A Queen of traffic jams and a Kingdom of order

An automobile society should be one with equal rights, where citizens are of good character, and follow the rule of law. People have become inured to traffic jams. I saw a WeChat comment on a Tianjin website a few days ago: "People's character is the key to fixing traffic jams, but they close off traffic control when the leaders drive about." That hit the target perfectly.

Let's talk about character. I've been going to Germany for more than ten years to test drive cars, not just German brands, but also new American, Italian, Korean, and Chinese models. The traffic regulations are the same in Germany as in China, but compliance is very different. German people stick to the rules.

For example, overtaking on the right isn't allowed in China but drivers do it anyway. Why is that? You have to overtake on the right because the overtaking lanes are full of slow cars. Chinese motorists don't care how fast or slow they drive. The fast lane is taken up by women who are still learning to drive and men talking loudly into their mobile phones, causing long lines of cars to form behind them. In Germany, cars going less than 120 kilometers per hour don't dare drive on the left. Even if a car is driving on the left, it will move to the right when it finds there is a faster car coming from behind.

Pedestrians are another example. In developed countries, pedestrians and drivers obey traffic signals. Pedestrians have the right of way on crossings without signal lights. Respect and understanding between pedestrians and vehicles make for good traffic order. The important thing is to obey the rules, even when others don't.

European drivers are not as impetuous as Chinese drivers, who cut in, argue, and even hit each other. Europeans drive peacefully. When you enter a main road from a side road, you must stop and yield to vehicles on the main road. That makes driving on the main road easier. When you see a car on a side road, you don't need to worry about preparing to brake, you can just step on the accelerator and speed by.

That orderliness makes for efficiency. I came to a road under construction in Germany. Cars had to merge into narrow lanes both ways. Cars were lined up on both sides, but no one blew their horns or pushed across the intersections bringing traffic to a halt. Instead they merged into the narrow lanes one by one in an orderly manner, like a zipper closing, making it much faster to get past. The Germans do everything methodically, and are not impatient. When they drive, they consider both their own convenience and the comfort of others. They take care to live their own lives without harming others. That makes their lives a virtuous cycle, and a paradise of order.

The Queen of England is supreme in Britain and in the world, but they don't close off lanes of traffic for her when she travels. Two police motorcycles usually clear the way and other cars give way. When the road is crowded, and the Queen's Rolls-Royce is blocked on the road, people driving alongside are delighted to recognize her car, and the Queen calmly waves to them through her window.

3. Some advice for the Chinese car industry

Protection is a bad thing

In the final section of this book, it's time to talk about visions of how the Chinese car industry can get bigger and stronger. But the future isn't something I can predict. I can only offer a few suggestions.

My calls since 2009 to drop 50-50 equity restrictions on joint ventures have caused an outcry. Critics have included auto industry officials, state-owned enterprises, and "patriotic" automotive media people. Geely Automobile Chairman Li Shufu is the only one who has openly supported me. The main thrust of their criticisms is nothing more than that China's auto industry is still in its infancy and should continue to be protected and supported, and China would be hit hard by liberalizing equity.

History has proven time and again that protection is a bad thing. It produces fools while competition creates strong competitors. China will never become an automotive power by relying on protection.

Observing Chinese cars for 40 years, I have realized what the draw-backs are.

Since its birth, the automobile industry had the nation's highest degree of protection. Imported cars underwent strict approval, and import duties of up to 220 percent were levied. As a result, until 1984, the engine power and overall standard of the Shanghai brand with an annual production of 2,000 or 3,000 vehicles, were not nearly as good as the copied Mercedes-Benz in the 1950s. The Red Flag with an annual output of less than 300 vehicles stopped production because of high fuel consumption and poor performance.

If Deng Xiaoping hadn't decreed that cars could enter into joint ventures in the early 1980s, no one would have been able to get past the block on foreign investment. By allowing joint ventures, the Chinese car industry was able to reach high international standards and reduce the gap by 20 years.

If indigenous brands hadn't been allowed in in 2001, indigenous innovation wouldn't be on the agenda today. If we hadn't joined the WTO, and competed and cooperated with multinational companies in globalized markets, becoming the world's largest automobile producer would have been unthinkable.

The 50:50 joint venture equity ratio originated in the interplay between Chinese and foreign parties. The future of joint ventures was uncertain in the 1980s, and we were feeling our way, each hoping for the other party to take on more responsibility and risk. The 50:50 ratio was arrived at when Shanghai Volkswagen was established. The foreigners couldn't survive without the support of the Chinese, whatever the share ratio was. They were few, and the Chinese were many. They were like mice avoiding being trampled on by elephants. The 50:50 ratio was later codified in the 1994 China Automotive Industry Policy, and was a clause that China insisted on in the WTO negotiations.

The Chinese industry and the global landscape have changed tremendously. It won't harm China to talk about liberalizing shares and even allowing foreign investors to set up wholly-owned companies.

Some suspect that foreigners want to expand their shareholding to strangle indigenous brands. People frighten themselves with such conspiracy theories. PSA Asia president Olivier Mornet supports lifting the restrictions on share ratios. His reasoning is that the 50:50 ratio is the least efficient for joint ventures. Decision-making needs repeated discussion on strategy formulation and causes friction. Market opportunities are thereby lost.

I asked Professor Jochem Heizmann, CEO of Volkswagen China, if Volkswagen had a strategy to strangle its own brands in China? The translator didn't want to translate my question, but I told him to. Professor Heizman stood up and said, "No! That has never been an option for Volkswagen."

Now that China has joined the WTO we're no longer in a domestic competition, but a "World Cup" where everyone is equal under international rules. We shouldn't let the Chinese team score without even letting foreign teams take a shot at the goal.

It wasn't until 2018 that there was some relaxing of the share ratio, with the official announcement it would be protected for another four years before being freed up. I don't think it would be terrible if there were 100 percent owned Volkswagen, Toyota, and Tesla in China. Indigenous brands like Geely, Great Wall, GAC Trumpchi, SAIC Roewe, Chang'an and Weilai will survive in full and fierce competition.

It's the logical way in a competitive global industry. If we can't do that, we should stop talking about being an automotive power.

We need craftsmanship, but systems are more important

Guo Qian was a senior manager at FAW, Beijing Hyundai, Volkswagen, and Chery. He talked about the difference between Chinese tailoring and the international garment industry, calling the former a craftsman culture and the latter a systems process.

He said that there was no lack of a craftsman spirit in Chinese tradition. The old tailors of the Shanghai Red Group at the end of the 19th century could make clothes that fit customers perfectly just by looking at them. Those craftsman methods met the basic need of making clothes for individuals, but they couldn't produce world-class mass-produced high-end garments. The old Chinese tailors understanding of fashion, quality, and fine workmanship were far from international standards. No matter how cheaply Chinese people were able to make suits, no one bought them internationally. If our understanding stops at cost and capability, we'll never reach international levels with the clothes and cars we make.

As Guo Qian sees it, a process systems culture is the combining and managing of countless amounts of data, and the classifying and matching of a lot of user needs, to obtain a result. Italy's name-brand garments can't be produced by one master, or even ten. They require labor by hundreds of people.

With a car brand, if you start out from experience, you are still like the tailors sizing up their customers. When you start using process systems management, you have to consider an infinite variety of combinations. There are a thousand kinds of buttons, to match with a thousand customers, and the correct results are brought to management.

Guo Qian opened his computer and showed me drawings of Chery and Mazda fuel tanks. There were only two or three drawings for Chery, but dozens for Mazda, with a lot of detailed data and text descriptions of processing standards. Mazda and Chery are far different in terms of process specifications. BMW is stronger than these, but there have been problems with BMW's fuel tanks and there have been recalls. When hundreds of thousands of cars are used in thousands of different environments, and there are hundreds of millions of combinations, ensuring quality is a matter of probability management. Even if the usage environments of millions of different users are simulated, it is difficult to guarantee that low probability problems won't occur. It's much more difficult than controlling the quality of a space shuttle. China's auto industry can't go forward unless it breaks free from its craftsman culture.

Guo Qian said, "It's like you are training a professional football team but they haven't touched the ball for a month. They manage to come up with some lucky shots. But it's hard to play a proper game when they keep making mistakes. When Toyota hired people for their Tianjin plant, they didn't want people who had car making experience. They just had to know Japanese, and were like a blank piece of paper, ready to start from the beginning."

If we keep going the way we did back when China first started making cars, we'll never reach the standards of Europe and the US. Our only hope is to change our strategy and goals and go the international way.

Guo Qian praised FAW for using the process systems model in developing the Pentium B50 from the B70. IBG was commissioned to develop the body of the B70. FAW engineers learned how to do it each step of the way, spending 50 million yuan. They commissioned IBG again to develop the B50, but this time they could do a lot of it themselves, and it only cost 25 million yuan.

IBG worked with almost every indigenous brand in China. Guo Qian asked IBG which Chinese company's development capability was nearest to international levels. They said they had not been in contact with SAIC; of the companies they were working with, only FAW was on the right track. Most indigenous brands had not mastered the basics of standardization. FAW's procedures for product development, product changes, preparation of product manifests, and effective alteration of standard drawings have been in place for 50 years. These basics were laid down when the Soviet Union helped build FAW, and they are still of great benefit.

When FAW started indigenous development ten years ago, it was difficult to develop competitive products according to standard processes. We were on the outside for 20 years, but craftmanship wouldn't get us to good international standard cars.

SAIC used Rover's 200 person R&D team to produce the Roewe 750. International standard R&D processes were adopted for the subsequent 550 and 350 and a lot of progress was made.

Having been able to analyze for themselves where the gap is, Geely and Chery are in transition. Roewe and Pentium are also bringing in a process management culture. In international competition where systems are all important, indigenous brands now have some hope.

Banning internal combustion vehicles shouldn't be an option

In September 2017, an official in a position of authority proposed at an automobile forum that as several European countries had recently announced plans to ban the sale of gasoline-fueled vehicles, China should consider a comprehensive ban on production and sales of traditional energy vehicles. To use a vulgar Chinese expression, a donkey should kick that fellow in the head.

As China is a major automotive country with the world's largest annual output of 30 million vehicles, a wise official action regarding the ban in those countries should be to formulate a strategic response, and not just follow them.

When someone hits you, don't hit yourself too.

Northern European countries like Norway which rigorously protect the environment don't produce cars themselves but import electric vehicles. Gasoline vehicles can't be sold, but that hasn't negatively affected their economies.

Britain used to have a magnificent car industry but it was basically sold off to German, Indian, and Chinese makers, so getting rid of gasoline vehicles doesn't matter to them.

Volvo announced it would stop selling conventional gasoline and diesel vehicles by 2019 and all models would be fitted with electric motors. But that was not saying that products couldn't also have gasoline engines. In fact, it was going the way of the plug-in hybrid.

Toyota proposes that by 2050, its carbon dioxide emissions will be reduced by 90 percent compared to 2010. Its products will include pure electric, hybrid, plug-in hybrid, and hydrogen fuel cell vehicles.

The Volkswagen Group has the most aggressive plan and will spend 50 billion euros. But at the same time, it is not relaxing efforts to optimize traditional internal combustion engines.

GM Chairman and CEO Mary Bora revealed that ten new energy vehicles will be launched in China by 2020. Although these include the Chevrolet Bolt pure electric vehicle, the Buick Velite extended range electric vehicle is still the main focus.

Technology changes every day, and there will be new sources of clean energy in the foreseeable future. These will certainly become the norm.

Is it realistic to set a timetable to end gasoline car travel for ordinary people in this country?

In my opinion, the timetables for bans on combustion engines by those European governments are actually intended to suit current politics. They are an indicator of the future but don't break new ground.

Mary Bora said, "GM is developing electric vehicles, but consumers, not the government, should decide what energy source they use. Let consumers choose the technology that meets their needs. Don't tell them what to do."

I am most concerned about coal and gasoline, the two major energy sources Chinese people currently have. Which will be better environmentally when both have been given the high-tech treatment? One should not replace the other.

Oil security is an important starting point for committing to electric vehicles. If electric and combustion vehicles develop in parallel, that security can be largely guaranteed. However, national security will be threatened if we move away from the use of gasoline.

Civilian transport infrastructure becomes strategic in wartime. It plays a key role in evacuating civilians, transporting wounded and sick people, and delivering military supplies during wars and major natural disasters.

In a regional war, or in a major earthquake, people become aware of the vulnerability of electricity production and distribution. Electric vehicles would be unable to refuel, thus paralyzing transportation networks. The short-sightedness of moving away from gasoline and the painful and unacceptable costs would become apparent.

This may be a truth that post-90s netizens understand.

A complete ban on combustible fuel shouldn't be an option.

Being an automotive power can't be just a daydream

In 2001, the year China joined the WTO, only 600,000 cars were produced and sold nationwide. There were more than 24.7 million in 2017. Of those, 10.85 million, or about half, were domestic Chinese brand vehicles.

We used to agonize about Chinese brands when they were in their infancy. However, they have emerged as a new force, with opportunities in SUVs, intelligent cars, and new energy vehicles. They even surpass multinational companies in design, quality, innovation, and price. And they are ordinary people's first choice.

Some indigenous brands have risen to the top and China is no longer in a position of weakness. Like aerospace, high-speed rail, and communications equipment, automobiles will make us a great power.

Three conditions have to be met for this to happen. There must be a number of well-known companies and brands with international competitiveness and international visibility. Domestic and international markets must be skillfully handled. And they must possess core technologies as well as insight into future trends.

How far are we from achieving these conditions? If I may be forgiven for speaking directly, we are deluding ourselves if we think we can do it in three to five years with slogans and an industrial leap forward.

Only the US, Germany, and Japan can be called automotive powers. Italy and France, with a history of famous cars and factories, are still not there. South Korea is still far from it despite its hard work. China has several decades of homework to do before its dream of being an automotive power comes true.

Wearing the crown as the world's largest automobile producer has made Chinese people extremely proud. However, "tangled" is the best way to describe the industry today. Bad practices such as shallow boastfulness, desire for quick profits, backward thinking, and structural weakness will make the dream hard to come true. A major issue for China in the next 60 years will be moving from a major automotive country to an automotive power.

Many signs indicate that as of 2018 the avalanche of growth has come to an end, and negative growth has occurred for the first time since the 2008 global financial crisis.

Purchasing restrictions in Beijing are just the first shot, and by 2018 almost all of the first-tier cities have followed suit. At the national level, the industry has lost its guaranteed growth status. Policies encouraging car consumption have come to an abrupt halt. Represented by the most stringent regulations in the world, emissions and quality standards have become stricter and harder to meet. Western countries have been pressing for intellectual property protection and open markets.

China lacks many things, but it has never lacked an eagerness for fame. The car market boom has swollen people's heads. Things cannot be believed when they are boasted about too much. The rising tide has concealed problems whose nakedness is only revealed at low tide.

After all, cars are a traditional industry, unlike new industries such as IT, which can explode overnight. The effects of capital operations are limited, and it will be technology and products that ultimately determine its destiny.

After ten years of ups and downs, Yin Tongyao, head of Chery, admitted, "Making cars is no differentfrom digging earthworks. It's a continuous function, every shovel counts, and you go up layer by layer."

I am anxious about the future of the Chinese automobile industry. Being first in the world makes my head spin and we've spread ourselves too thin. It's like trying to put three lids on five teapots. If the environment changes suddenly or the market cools, the funding chain will be broken. I'm afraid hard-earned family property will suffer a big hit in two or three years and that fear is not unfounded. Just look at the ups and downs of the Big Three US manufacturers over the past ten years. Becoming an automotive power can't be just a daydream like in Feng Xiaogang's film.

"People will always love cars, but we have forgotten how to show our love."

The luckiest part of my career is that I have been able to stand in a unique position for 40 years, and witness ordinary Chinese people breaking through barriers and gaining the right to a car culture. China joining the WTO opened a gate that we can't close. I hope that in the next 60-year cycle, we take advantage of competition pressures and standardized rules to make positive changes in the industry. We must establish a culture that is compatible with advanced contemporary productive forces, have a rational, open, tolerant, and humble mindset, and ride the tide of world change.